# PRELUDE TO PHYSICS

## CLIFFORD SWARTZ
State University of New York at Stony Brook

**JOHN WILEY & SONS**
New York   Chichester   Brisbane   Toronto   Singapore

Copyright © 1983, by John Wiley & Sons, Inc.

All rights reserved. Published simultaneously in Canada.

Reproduction or translation of any part of
this work beyond that permitted by Sections
107 and 108 of the 1976 United States Copyright
Act without the permission of the copyright
owner is unlawful. Requests for permission
or further information should be addressed to
the Permissions Department, John Wiley & Sons.

*Library of Congress Cataloging in Publication Data:*

Swartz, Clifford E.
 Prelude to physics.

Includes indexes.
1. Physics.   I. Title.

QC21.2.S913   1982   510   82-16037
ISBN 0-471-06028-3

Printed in the United States of America

10 9 8 7 6 5 4 3 2

# PREFACE

This is not a physics text, nor is it a math review book. It is a prelude to physics. We assume that you are about to take, or have already started, a regular course in college physics. For some people, that course is no fun. For many others, it presents a hurdle that must be overcome. Worse yet, sometimes the hurdle turns into an absolute roadblock.

Often the hurdle is needlessly high. The work comes too fast and is too different from high school science. What we have done in this book is to gather together the most common stumbling blocks experienced by students of the introductory physics course. It may appear that many of the problems are mathematical. Indeed, the physics course makes constant use of various branches of mathematics. But the mathematics needed is a kind seldom emphasized in math courses, and rightly so. Mathematicians are concerned with rigorous proofs and generalized cases. Physicists and engineers are more concerned with how to use math as a tool to describe phenomena. Instead of rigorous proofs, we want plausibility arguments. Instead of generalized cases, we want specific applications, usually with numbers attached. In a math course, for example, the area of a circle with a radius of 2 is equal to $\pi r^2 = 4\pi$. For our purposes, such a statement is meaningless, and thus useless. The radius must have units attached: 2 m. Since the value of the radius must have been determined by a measurement, we must specify the precision of that measurement in terms of the significant figures of the number. Is the radius 2 m or 2.00 m? Finally, for our purposes, the area of the circle is about 12 m$^2$, not about $4\pi$ m$^2$. Do not ever try to order a piece of metal with an area of $4\pi$ m$^2$ from a stockroom clerk!

Some people think that physics is the science of precision. Nonsense. Physics is the science of common sense. In solving problems we will zero in on the solution, figuring out a first approximation, and then a second or a third, if that is necessary. In the first sets of lessons we will learn how to estimate orders of magnitude, and then how to specify greater precision if there is need to do so.

After learning how to specify and manipulate numerical quantities, we will investigate and measure a number of phenomena that can be described as simple mathematical functions. Most of the phenomena of the first-year physics course can be described with only four such functions—the power function, the sinusoidal function, the exponential function, and the logarithmic function. In the process of describing these phenomena and using these functions, we must learn how to deal with the functions algebraically and graphically.

All of the new introductory physics texts use the SI (Système International) units. This is a metric system with names established by international agreements. We must become familiar with these units, not only with how the symbols appear on paper, but with the magnitudes as felt by our muscles, or in terms of other daily experience.

Phenomena take place in space. If we are to describe the phenomena we must be able to do so in terms of geometrical spaces and coordinate systems. Therefore, we will study vectors and analytical geometry, but always in terms of phenomena that will later be studied in the physics class.

It may sound as if we are going to be grubbing about among the foundations of physics. We will be down there, all right, but it turns out that some of the most interesting and current problems of modern physics are not far removed from its foundations. So while we learn how to measure your foot length, we will also consider how to measure the diameter of an atom. While we learn how to measure an angle on a table top, we will also show you how you can easily measure for yourself the circumference of the earth. One of the pleasures of studying physics is that some of the simplest tools can be used to determine relationships that are profound.

**Clifford Swartz**

# CONTENTS

| | | |
|---|---|---|
| CHAPTER 1 | HOW TO WRITE NUMBERS, LARGE OR SMALL | 1 |
| CHAPTER 2 | FERMI QUESTIONS AND ORDER-OF-MAGNITUDE | 5 |
| CHAPTER 3 | PRECISION AND ERROR | 9 |
| CHAPTER 4 | UNITS AND MEASUREMENT OF MASS | 27 |
| CHAPTER 5 | THE SIMPLE MEASUREMENT OF TIME | 30 |
| CHAPTER 6 | UNITS AND DIMENSIONS | 34 |
| CHAPTER 7 | SOME GENERAL COMMENTS ON GRAPHS | 37 |
| CHAPTER 8 | TRIGONOMETRY AND USEFUL ANGLES | 46 |
| CHAPTER 9 | GEOMETRY AND DIAGRAMS | 59 |
| CHAPTER 10 | MOTION AND THE POWER FUNCTIONS: $y = kx^n$ | 66 |
| CHAPTER 11 | FORCE AND NEWTON'S SECOND LAW | 104 |
| CHAPTER 12 | VECTOR ADDITION | 111 |
| CHAPTER 13 | MOMENTUM | 123 |
| CHAPTER 14 | ENERGY AND THE MULTIPLICATION OF VECTORS | 132 |
| CHAPTER 15 | THE SINUSOIDAL FUNCTION: $y = A \sin \theta$ | 145 |
| CHAPTER 16 | THE EXPONENTIAL FUNCTION: $y = a^x$ | 160 |
| CHAPTER 17 | LOGARITHMS: $y = \log_a x$ | 172 |
| APPENDIX 1 | USEFUL FORMULAS AND RELATIONSHIPS | 192 |
| APPENDIX 2 | GREEK LETTERS AND THEIR COMMON USE IN PHYSICS | 195 |
| APPENDIX 3 | THE INTERNATIONAL SYSTEM OF UNITS (SI) | 198 |
| | INDEX | 201 |

# CHAPTER 1
# HOW TO WRITE NUMBERS, LARGE OR SMALL

The radius of a proton is 0.000,000,000,000,001 meters. The age of the universe is 500,000,000,000,000,000 seconds. Clearly, this is no way to describe the physical universe. It's hard to keep track of all those zeros. There is a better scheme, called in the schools, unfortunately, "scientific notation." Since that name is unknown in the outside world, we shall simply call the method the "power-of-ten" notation. Here's a translation of the notation:

$$\begin{array}{ccccc} 0.01 & 0.1 & 1 & 10 & 100 \\ 10^{-2} & 10^{-1} & 10^{0} & 10^{1} & 10^{2} \end{array}$$

Some of the symbols have common names. $10^{-1}$, for instance, is one-tenth. $10^2$ is a hundred. $10^3$ is a thousand. $10^6$ is a million. $10^9$ is a billion. It's obvious why $10^2$ is 100, since $10^2$ means ten squared. $10^1$ is just ten to the first power, which is 10. It may not be so obvious why $10^0$ is equal to 1. How can you raise any number to the zeroth power? The explanation lies in the multiplication properties of numbers written in this way.

Let's multiply 1/100 by 1000:

$$\frac{1}{100} \times 1000 = 10$$

In the power-of-ten notation, this equation is written

$$10^{-2} \times 10^{3} = 10^{1}$$

When multiplying numbers written in this fashion, you add the exponents: $-2 + 3 = +1$.

Note that a negative exponent merely means the reciprocal of the number. For example, $10^{-3} = 1/10^3$. Suppose that we apply this rule to the product of 1/100 and 100:

$$10^{-2} \times 10^{2} = 10^{-2+2} = 10^{0} = 1$$

We would get this same result regardless of the exponent: $10^{-a} \times 10^{a} = 10^{-a+a} = 10^{0} = 1$.

With the power-of-ten notation we can write the radius of the proton as $1 \times 10^{-15}$ meters (m). The age of the universe is $5 \times 10^{17}$ seconds (s). Note that we have now added one digit before the power of 10. We can write any number this way. For instance, there are $3.65 \times 10^2$ days in a year (yr). There are $3.600 \times 10^3$ s in 1 hour (h). It is not necessary to put just one significant figure to the left of the decimal point, but it is frequently convenient to do so.

The power-of-ten notation is a convenient way of writing numbers, but it also provides a convenient way of multiplying and dividing them. Suppose that you want to multiply 8346 and 2683. Of course, you can do it longhand, but these days you would be more apt to turn on your hand calculator and punch in the numbers. Sometimes, however, you don't have your hand calculator or its batteries have run down. There is an easy way to get a good first approximation, either in your head or on the back of an envelope:

$$8346 = 8.346 \times 10^3$$
$$\times \underline{2683} = \underline{2.683 \times 10^3}$$

We can multiply separately the digits and the powers of ten. As for the digits, note that we are multiplying a number a little less than $8\frac{1}{2}$ by a number a little larger than $2\frac{1}{2}$. That's going to yield a number a little larger than 20. Meanwhile, the power-of-ten product is $1000 \times 1000$, or 1,000,000. Our answer, then, must be a little larger than $20 \times 10^6$, which could also be written $2 \times 10^7$. For many purposes we have no need to know the answer more precisely. If you do have to know the answer more precisely, you can find your calculator, charge up the batteries, or, if worse comes to worst, do the problem longhand.

> At least, with this method, you have a first approximation that is easy to calculate, and that can be checked against the answer derived in some more sophisticated method.

Never do a problem only one way! Throughout this book we will emphasize the point that problems should be attacked in several ways, usually by finding a first approximation or a "ball-park solution," and then using a fancier method. Frequently, the quick and easy approximation serves not only as a check on your final answer, but also as a guide to how to get it.

Here's another example of multiplying numbers with the power-of-ten notation. How many seconds have you lived? Suppose that you are 18 years old. Then the number of seconds you have lived is

$$18 \text{ yr} \times 365 \text{ days/yr} \times 24 \text{ h/day} \times 60 \text{ min/h} \times 60 \text{ s/min}$$

In power-of-ten notation, the numbers can be written

$$1.8 \times 10^1 \times 3.65 \times 10^2 \times 2.4 \times 10^1 \times 6.0 \times 10^1 \times 6.0 \times 10^1$$

You can do the digit multiplication in your head, something like this: a number a little less than 2 times a number a little larger than $3\frac{1}{2}$ equals 7. 7 times a number a little less than $2\frac{1}{2}$ equals 17. 17 times 6 is approximately equal to 100. 100 times 6 is equal to 600. So the digits multiply out to about 600, and the exponents in the powers of 10 add up to 6. Therefore, you have lived about $600 \times 10^6$ s, which might be rewritten as $6 \times 10^8$ s. If the method seems sloppy, and if you care enough about the precise answer to spend some of your remaining seconds, feel free to calculate it longhand.

Here's another example that combines both multiplication and division. This is just an arithmetic exercise—no units are used:

$$\frac{84.6 \times 0.0064}{9832 \times 8.6 \times 10^{-4}} \rightarrow \frac{8.46 \times 10^1 \times 6.4 \times 10^{-3}}{9.832 \times 10^3 \times 8.6 \times 10^{-4}}$$

Operating first on the digits, we can cancel out the 8.46 in the numerator with the 8.6 in the denominator. What's left is $6.4/9.8$ which approximately equals 0.6. Adding the exponents in the numerator (and subtracting the exponents in the denominator) yields $10^{-2}/10^{-1}$ equals $10^{-1}$. Therefore, the answer is

$$0.6 \times 10^{-1} = 6 \times 10^{-2}$$

> In this last problem, note the way that we added the exponents for multiplication, but *subtracted* the final exponent in the denominator because we are dividing.

Here are some other examples:

$$\frac{10^5}{10^3} = 10^2 \qquad 1\% = 0.01 = 1 \times 10^{-2} \qquad \frac{10^{-3}}{10^{-1}} = 10^{-3-(-1)} = 10^{-2}$$

$$\frac{10^5}{10^{-2}} = 10^{5-(-2)} = 10^7 \qquad 50\% = 0.5 = 5 \times 10^{-1} \qquad 0.01\% = 0.0001 = 1 \times 10^{-4}$$

## PROBLEMS (WITH ANSWERS)

1. $\dfrac{14{,}000}{383 \times 1.0\%} = $ _____ $\left(\approx 3\frac{1}{2} \times 10^3 = 3.7 \times 10^3\right)$

2. $\dfrac{0.084 \times 86}{9.2 \times 74} = $ _____ $\left(\approx \dfrac{70}{68} \times 10^{-2} = 1.1 \times 10^{-2}\right)$

3. $\dfrac{4623 \times 98 \times 41}{0.0062 \times 122} = $ ⬚  $(2.5 \times 10^7)$

4. $\dfrac{18}{6 \times 10^{23}} = $ ⬚  $(3 \times 10^{-23})$

5. $\dfrac{10{,}001 \times 0.08}{143 \times 9.87} = $ ⬚  $(0.6)$

# CHAPTER 2
# FERMI QUESTIONS AND ORDER OF MAGNITUDE

One of the great physicists of this century was Enrico Fermi (1901–1954). He not only made major contributions to theory but was also a skilled experimentalist. He was legendary for being able to figure out things in his head, using information that seemed too sparse for any quantitative conclusion. He zeroed in on problems by saying that a particular value surely must be less than a certain amount but larger than some other figure. In the end he would have a quantitative answer that must surely have been correct within certain limits. For example, one of his most famous calculations on a nonscientific topic concerned the number of piano tuners in New York City. Off hand, how could one possibly figure out such a thing? Here's one way. The number of piano tuners in a region must be connected in some way with the number of pianos, and that, in turn, depends on the number of people in the region. In what follows, we'll make some assumptions about certain factors concerning population and piano ownership. If you don't like our assumptions, try your own. There must be about $1 \times 10^7$ people in New York City. People don't own pianos, however; families do. Perhaps there are $2 \times 10^6$ families in New York City. Certainly not every family owns a piano, but perhaps one out of five does. In that case, there are $4 \times 10^5$ pianos in New York City. Some people tune their pianos every month. Some don't tune them for years. If we assume that on the average every piano gets tuned once a year, then there are $4 \times 10^5$ piano tunings per year in New York City. How many pianos can a piano tuner tune per year? Some might tune as many as four per day. If we assume that there are 200 working days per year, then every piano tuner could tune 800 pianos per year. The number of tuners needed is, therefore, $\frac{4 \times 10^5 \text{ tunings per year needed}}{800 \text{ tunings per year per tuner}} = 500$ piano tuners. You see, we did know how many piano tuners there are in New York City! You can try different assumptions for our various factors, if you wish, but it is highly unlikely that you can justify an answer greater than a factor of 10 or smaller than a factor of 10 from the number that we have obtained. That is, it is unlikely that there are more than 5000 piano tuners in New York City or fewer than 50. We have obtained an answer that is probably good to within an "order of magnitude."

Physicists and engineers frequently describe quantities in terms of a "factor of 10" or a "factor of 2." These expressions mean that you are multiplying or dividing by the factor. The most common meaning for "an order of magnitude" is a factor of 10. Our power-of-ten notation automatically gives the order of magnitude of numbers in terms of the exponent of the 10. If an auditorium has 1000 seats and is nearly full, the order of magnitude of the people in the hall is $10^3$. If ten more people walk in or walk out, the order of magnitude does not change. If half the people walk out, the audience has decreased by a factor of 2.

Physicists and engineers, as well as economists, use the Fermi technique in their everyday work. You use the same technique when you go shopping. If you are going to buy groceries you probably know the money that you have available to within a few percent. As you pick items off the shelves you are probably keeping a rough track of the sum to make sure that you have enough money to cover your purchases. You do not start out your week's grocery shopping with a $1 bill or with a $1000 bill. Before you start, you know the order of magnitude of the amount that you are going to spend.

Before tackling the exercises on the next page, consider how much you really know about the sizes of things. For instance, how high is a two-story house? On each floor you could almost touch the ceiling by standing on your tiptoes. Therefore, the height must be between 7 and 8 feet (ft), or about $2\frac{1}{2}$ m. (The standard height of rooms in most American houses is 8 ft.) The two-story house must therefore be at least 5 m high, even with a flat roof, and with a normal peaked roof may rise another 2 m. So the top of a two-story house must be about 7 m above the ground, which is about 21 ft. Add a foot or two if the first floor is slightly above ground level. At the very least we are safe in claiming that a two-story house is at least 5 m high and no more than 10 m high. With a little bit of nerve you can make similar arguments about many other quantities. You should not only produce an estimate of the expected size, but also the bounds within which the measured value must surely lie.

## FERMI PROBLEMS

1. What is the mass in kilograms (kg) of the student body in your school? [1 kg weighs 2.2 pounds (lb).]

2. What is your density in kilograms/cubic meter (kg/m$^3$)? (If you know your weight in pounds you can find your mass in kilograms. As for your volume, what size cylinder could you just stand in? The volume of a cylinder is $\pi r^2 h$. To get the volume in cubic meters, you must measure your radius and height in meters. When you complete the calculation, consider that the density of water is 1000 kg/m$^3$. Since you probably just about float in water, your density should be about the same as that of water.

3. How many golf balls will fit in a suitcase? In this case, you may want to estimate the length, width, and height of some typical suitcase in inches (in.). Assume that each golf ball is a 1 in. diameter sphere. The volume of a sphere is $\frac{4}{3}\pi r^3$. However you can't fill up a suitcase with golf balls leaving no spaces inbetween. It would be close enough to assume that the volume of a golf ball is 1 in.$^3$. Note what happens to your final answer if you assume that the diameter of the golf ball is 2 in. The volume would increase by a factor of $2^3 = 8$, almost an order of magnitude.

4. How many cells are there in the human body? This question again illustrates how some information can be obtained out of very little definite knowledge. Living cells come in a great range of sizes. However, they can be seen with an ordinary light microscope, and therefore must have a diameter larger than the wavelength of light. They can scarcely be seen with the unaided eye and so must have a diameter smaller than 0.1 millimeter (mm) = 100 microns ($\mu$m). A reasonable assumption would be that the average cell diameter is 10 $\mu$m = $10^{-5}$ m. If you assume that cells are spherical, you can calculate the volume of the average cell. Use your estimated value for the volume of a human and figure out the number of cells in the human body.

5. How many hairs are on your head? Assume some reasonable spacing between hairs—perhaps 1 mm. With this assumption, there would be 10 hairs per centimeter or 100 per square centimeter (cm$^2$). You will have to figure out the approximate area covered by the hair on your head.

6. How many individual frames of film are needed for a feature-length film? How long is such a film? To work on this problem you must know that motion pictures consist of a series of still pictures presented at a rate of about 24 per second. With 35 millimeter (mm) film, the width is 35 mm and the height is 24 mm, or about 1 in.

7. What is the ratio of molecular diameter to spacing between gas molecules in a gas at standard temperature and pressure (S.T.P.)? A mole ($6 \times 10^{23}$) of gas molecules at S.T.P. occupies 22.4 liters (l). A molecular diameter might be $2 \times 10^{-8}$ cm. Compare available volume per molecule with the volume of a molecule.

8. If your life earnings were to be doled out to you at a certain rate per hour for every hour of your life, how much is your time worth?

9. What is the weight of solid garbage thrown away by American families each year?

10. How many molecules are in a standard classroom?

# CHAPTER 3
# PRECISION AND ERROR

We have all known some significant figures in our lifetime, and if you have ever seen data presented in newspapers or other popular accounts, you have probably seen figures that were insignificant, if not outright lies. *Significant figures* is a technical term describing the meaning of digits in numbers. Whether or not there are really five significant figures in the number 32,584 meters depends on the reliability of the person who made the measurement and whether or not he or she was correctly using the convention. There is indeed an accepted convention for the use of significant figures. The common understanding is that the last digit in a number is guaranteed to be valid within one-half unit. For instance:

$$2.5 < 3 < 3.5 \text{ (2.5 is less than 3, which is less than 3.5)} \qquad 2.95 < 3.0 < 3.05$$

The 3 all by itself is just one significant figure. If we say that there are three people in a room, we certainly do not mean that there are more than $2\frac{1}{2}$ but less than $3\frac{1}{2}$; we mean that there are exactly three people in the room. Still, if we say that the height of the room is 3 m, according to the convention, we are guaranteeing only that it is between 2.5 and 3.5 m. The number 3.0 has two significant figures. Even though the second digit is a zero it contains the guarantee that our knowledge of the number is precise within plus or minus one-half of that final digit, in this case within one-half of 1/10.

The number 1.00001 has six significant figures. On the other hand, the number 0.00001 has only one significant figure. This last number, for instance, could be written $1 \times 10^{-5}$.

Frequently, the use of significant figures to represent measured data is not very satisfactory. You might measure the length of a line to be 6.1 cm, thus guaranteeing that the length is larger than 6.05 but less than 6.15 cm. You are claiming that you know the length of the line to within + or − $\frac{1}{2}$ mm. If you were using a good machinist's caliper, you might well know the length to within + or

− 2/10 mm. Another common possibility is that you could guarantee that the length could not be less than 1/10 mm short of 6.1 cm, but might be larger by $\frac{1}{2}$ mm. The use of significant figures does not provide a precise way of describing precision. Using significant figures correctly is something like using good grammar. It is something that one simply must do. However you must also go beyond the convention. Let's consider the use of significant figures to be a first approximation in describing the precision of data. In the next section we will see a much more versatile second approximation.

There are conventional rules for keeping significant figures straight during addition and multiplication. For addition or subtraction, the sum or difference has significant figures only in the decimal places where *both* of the original numbers had significant figures. The rule is best illustrated with examples:

$$
\begin{array}{cccccc}
5.2 & 6.843 & 6.843 & 6.843 & 500 & 5.00 \times 10^2 \\
3.1 & 1.2 & 1 & 0.001 & -\ 4 & -\ 4 \\
\hline
8.3 & 8.0 & 8 & 6.844 & 500 & 4.96 \times 10^2
\end{array}
$$

Note that 0.001 has only one significant figure, yet when added to 6.843 the answer has four significant figures. The sum can have more significant figures than one of the original numbers. It is the decimal *place* of the significant figure that is important in addition and subtraction.

Note the ambiguity of a number written as 500. Are we guaranteeing that the true number is between 450 and 550? Or is it the following case: 499.5 < 500 < 500.5? Evidently the zeros in 500 are confusing. It would be better to write such a number in power-of-ten notation. Perhaps the number is $5 \times 10^2$, which has one significant figure; or perhaps it is $5.00 \times 10^2$, in which case there are three significant figures. Both cases are illustrated in our sample additions and subtractions. For instance, if there are 500 people in an auditorium, and four leave, there are still 500 people in the auditorium, assuming that by 500 we mean $5 \times 10^2$. On the other hand, if you have $5.00 \times 10^2$ dollars in your bank account, and take out 4 dollars, you have $4.96 \times 10^2$ dollars left.

In multiplication and division, the product or quotient cannot have more significant figures than there are in the least accurately known of the original numbers. Once again, the rule is best understood in terms of examples:

$$
\begin{array}{r}
5.2 \\
\times\, 3.1 \\
\hline
16.12 \to 16
\end{array}
$$

Since each of the original numbers has only two significant figures, the product can have only two significant figures. The rule is not completely arbitrary. Consider, for example, how large or how small the product might be if we took maximum or minimum values for the individual numbers:

$$\begin{array}{cc} 5.25 & 5.15 \\ \times 3.15 & \times 3.05 \\ \hline 16.5375 & 15.7075 \end{array}$$

According to our standard convention for significant figures, the product can lie between 15.7075 and 16.5375. Therefore, we are justified in claiming that the answer is 16, which, according to the convention, means between 15.5 and 16.5.

Here are some other examples of multiplications and divisions making proper use of significant figures:

$$5.243 \times 3.1 = 16$$

One of these numbers has only two significant figures. Therefore, the product cannot have more than two significant figures:

$$5.243 \times 0.0031 = 0.016$$
$$37/9 = 4 \qquad 37/9.1 = 4.1$$

There are many small inconsistencies in the conventional rules for significant figures. Here's an example. Suppose that you measure the length of a box to be 3.025 cm, the width, 2.5 cm, and the height, 2 cm. The straight arithmetical multiplication yields an answer a little larger than 15. According to the rules of significant figures, however, there should be only one significant figure in the answer, since the height is known to only one significant figure. If we followed this rule, we would have to claim that the volume of the box is 20 cm$^3$, the guarantee being that it is between 15 and 25 cm$^3$. We actually know the volume closer than that, however, as you can see for yourself by taking the maximum and minimum possibilities for the three figures. The maximum value is only a little more than 19, and the minimum value is about 11. In this case, it would be better to give the answer as 15 cm$^3$, even though the usual guarantee is not valid. In general, when a number ends with the digit 1, it is legitimate to use an extra significant figure. The fussiness and inconsistency of this special rule points up the need for a more precise way of specifying and dealing with error limits on numbers.

We are about to present some exercises in writing and manipulating significant figures. After you do the exercises, there may be a great temptation to think that the lesson is done, and you need not bother with them anymore. Remember, however, that the proper use of significant figures is part of the grammar of quantitative science. You must be especially careful to observe the grammar when using a digital calculator. When those flashing red numbers present the answer to you in ten glowing digits, *remember that it lies*. Round off those insignificant digits and *tell the truth*.

1. $2.000 + 0.01 = 2.01$
2. $2.000 \times 0.01 = 0.02$

(Justify the answers in 1 and 2 by calculating the extremes in the sum and product that could be justified by the extremes in the original numbers.)

3. $\dfrac{4832 \times 0.165}{15 \times 264} =$

4. What is the volume of a piece of chalk that is 10.5 cm long and has a diameter of 1 cm?

5. The total weight of 42 people is 6342 lb. What is the average weight? Now suppose that you have a table with an area of 6342 cm². The width is 42 cm. What is the length? (Look out! The numerical answers to these two problems are not the same. Why not?)

## ABSOLUTE AND PERCENTAGE ERRORS: A SECOND APPROXIMATION TO ERROR ANALYSIS

Note a strange thing about the title of this section. We will be talking about dealing with precision measurements, and yet we are also assuming there are errors. Indeed, there are always errors associated with measurements. (There is an exception. If the measurement consists of counting quantized quantities, such as the number of people in a room, then the number can be exact.) By "error," we do not at all mean mistake. The technical meaning of error might better be described by the word "uncertainty." In any measurement of continuous variables, such as length, mass, time, and so forth, there is bound to be uncertainty as to the exact value of the answer. The best we can do is to give a particular measured value with a guarantee that the "true" value is the same, plus or minus a certain small amount.

Here's an example of a measurement with errors. What is the width of this page? The length of 1 cm is shown by the distance between these two lines. |↔| Glance at that length and then glance at the width of the page and write down your estimate for the width of the page in centimenters: ▭ . Your measurement surely has some uncertainty about it—an error—but it's unlikely that you will be off by more than 50%. Estimate now what your error or uncertainty is. Can you guarantee that the width of the page is less than ▭ cm, but is surely more than ▭ cm? Suppose that you estimated that the page was 20 cm wide and was surely larger than 15 cm but less than 30 cm. You should then write the width of the page as $20^{+10}_{-5}$ cm.

It may seem to you that this procedure was not a measurement but simply a guess. Perhaps the way to make a real measurement would be to get a centimeter ruler and put it on the page. Do that and write the answer in the following blank: _____ . Is this measurement without error? Are you guaranteeing that your measurement is correct within $\pm 1$ cm? $\pm 1$ mm? $\pm 0.1$ mm? At some point don't you have to worry about the reliability of your measuring instrument? At some point, don't you also have to worry about the definition of the edge of the page on the side where it is bound?

Shouldn't you, at least in physics class, always try to reduce the uncertainty of your measurement, and thus have the minimum error? Suppose that you want to measure the area of your front yard. If your purpose is to buy lime for the grass, then the proper way to measure the area is to glance at it as you drive down the driveway on your way to the hardware store. Lime comes in 50 lb bags, is cheap, and the amount that you put on the grass is not critical. Suppose, however, that you want to buy some Kentucky Bluegrass seed to put on your lawn. You buy that seed by the pound, and it's fairly expensive. Too much, and the lawn will choke. Too little, and the grass will be scraggly. In this case, the proper way to measure the area of your front lawn is to "pace" it. You certainly wouldn't use a meter stick. The measuring pace is the double-step, a technique well known to all outdoorsmen and carpet sellers. The pace is a very ancient measure of length, used at least as far back as the Egyptians when they built the pyramids. For an adult male, the double-step is about 5 ft. Our word "mile" comes from the Latin "mille," meaning 1000. 1000 paces, or 1000 double steps, is about equal to our present-day mile. To measure the area of your front lawn, you would pace out the length and the width and perhaps draw a crude diagram if the area were not exactly rectangular. Suppose, however, that you have to know the area of your front lawn in order to pay taxes. Now you need the equivalent of the meter stick, which, for this purpose, turns out to be surveying instruments.

Each of these three methods of measuring the area of your lawn is appropriate for the task and need at hand. Any one of them would be inappropriate for either of the other two tasks. *Never make a measurement without knowing the purpose for which the information is needed.* If you do not know the purpose, then you do not know what precision is needed, and you do not know what instruments to use. Precision is usually expensive, either in time or money. It should never be obtained when it is not necessary. There is an old folk saying that "if a thing is worth doing, it is worth doing well." That's utter nonsense. In everyday life, as well as in the physics laboratory, "if a thing is worth doing, it's worth doing well enough for the purpose at hand." That philosophy won't necessarily make your life easier, however. It means that before you start a project you have to know what your purpose is. There are times when you don't have to worry whether you know the amount of money in your pocket closer than ±50%. There are other times when you ought to worry whether contaminants in your drinking water are greater than one part per million.

There's another very common misunderstanding about "error" in student measurements. Frequently, the word is used to mean the difference between the "truth" and the student's measurement. The "truth" presumably has been obtained from the teacher or from a handbook. But of course, there is an uncertainty, or an error, attached to the handbook value. The only proper way to compare your answer to a handbook answer is to see whether or not they agree within the overlap of their mutual errors. For instance, suppose that you are measuring the acceleration due to gravity $g$. To two significant figures, $g = 9.8$ m/s$^2$. According to our convention for significant figures, we have $9.75 < g < 9.85$. Of course, better precision can be obtained from handbooks, although the value does depend on the position on earth. Suppose with some measurement that you make in the laboratory you get an answer of $9.6 \pm 0.3$ m/s$^2$. You didn't get the same answer that the handbook did! Is your error $0.2$ m/s$^2$? Certainly not! Your answer, with its margins of error, overlaps the handbook answer, with its margins of error. No excuses are needed. You may want to explain how you determined the size of your uncertainties, and we will soon get into that question. What if your answer did not agree with the handbook answer within the overlap of the errors? Then you have some explaining to do. Perhaps you were too optimistic about judging your uncertainties. Perhaps you made a mistake. Errors are not mistakes. While we are going to develop suggestions and rules for dealing with errors, we can offer very little help for dealing with mistakes. The main way to guard against mistakes is to do each problem or each measurement in two different ways, or at least at two different times. Regardless of precautions, anyone can make mistakes.

We have been talking about precision, but have not mentioned the word accuracy. Are they the same thing? Centimeter rulers have been printed at the top and the bottom of the opposite page. The one at the top was drawn with great precision. The subdivisions go down to millimeters. At the bottom of the page, the ruler was sloppily drawn. It has only centimeter markings, with no subdivisions. Measure the page using each ruler, and record your results in the blanks.

Precision ruler [ ]     Sloppy ruler [ ]

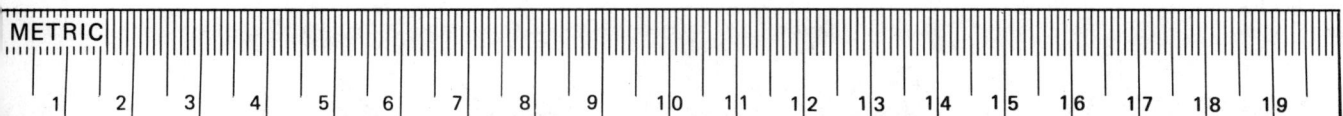

Surely you were able to obtain at least one more significant figure using the precision ruler than you could with the sloppy ruler. That extra significant figure, down to a fraction of a millimeter, makes the measurement look precise. How come, though, it is so different from the width measured by the sloppy ruler? Get a regular centimeter ruler and measure the width of the page. Consider the moral of the story, and you will know the difference between precision and accuracy. In the real world of measuring, you should never trust too blindly in the alleged accuracy of some instrument. Somebody else may have dropped it and put it back on the shelf just before you got it.

## SOURCES OF ERRORS

There are no general rules concerning the amount of error in a measurement. The experimenter has to use judgment, common sense, and experience in assigning reasonable error limits. There are some folklore rules—usually wrong. For instance, sometimes students are told that they should always read an instrument to the nearest $\frac{1}{2}$, or the nearest $\frac{1}{5}$, or the nearest $\frac{1}{10}$ of the smallest subdivision. No such rule can be general. Whether or not you can estimate values between the subdivisions on a scale depends on the type of instrument and your own experience. Sometimes school students are told that if they measure something ten times and then take the average, they will get one extra significant figure. There is no justification for this rule. If you measure the same thing ten times in a row, you will probably make the same biased observation each time.

1. Errors may depend on the technique with which you use a measuring instrument. For instance, Fig. 3-1 shows one way to read the width of a table with a meter stick. This arrangement is subject to *parallax*. The divisions on the meter stick are not immediately adjacent to the object being measured. As shown in the diagram, the reading depends on the relative position of your eyes as you look past

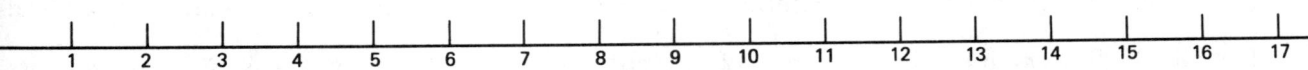

**Fig. 3-1**

the divisions to the object. Of course, you can eliminate parallax by sighting down perpendicular to the meter stick. However sometimes that's hard to do. Some electrical meters contain a mirror underneath the scale. You should line up the meter needle with its image in the mirror when noting the position against the scale. When you see that the needle and its image are aligned, then you know that you are looking perpendicular to the scale. An even simpler way to eliminate parallax with a meter stick is to turn it sideways so that the markings lie directly on top of the object being measured. There are similar problems of technique with almost every measuring instrument. With a pan scale, for instance, the measurement should be taken while the pans are oscillating slightly about the equilibrium position, not while they are at rest. Only experience with the actual instruments can teach you the proper way to use them.

2. Precision and accuracy, of course, depend on the nature of the measuring instrument. We have already seen this in measuring the width of the book. If you want to measure the diameter of a cylinder, you could be more precise, and probably more accurate, by using a machinist's caliper than a meter stick.

3. The precision to be obtained depends on the measurer's time and skill and prejudices. This point is different from just knowing the technique. You may very well know how to use a machinist's caliper to measure the diameter of a cylinder but may not want to take the time to get one. Certainly if you have no need for great precision, and it would take time to get a caliper, then a meter stick might do. In some cases, it may work just the other way around. A person may be so skilled and have precision instruments so ready at hand that he can make a precise measurement with no more trouble and no more time than he could a crude measurement.

4. The precision to be used in measurement also depends on the object being measured and the model of that object being assumed. For example, suppose that you want to measure the volume of a piece of chalk. If you assume that the shape of the chalk is a cylinder, then you should measure its diameter and height, and use the formula for the volume of a cylinder. Then if you had a machinist's caliper handy, you might be able to measure the diameter with a precision of $\pm 0.001$

cm. You would be kidding yourself, however. No ordinary piece of chalk is that round. The mathematical model of a cylinder is only a crude approximation.

## EXERCISES IN MEASUREMENT AND ERROR JUDGMENT

One of our purposes in this study is to make the SI units familiar. The best way to do that is to use these units over and over again in measuring familiar objects. The unit of length is the meter, about the size of a yard. The meter was originally chosen as a unit of length during the French Revolution. It was defined as one ten-millionth of one-quarter of the earth's circumference. Surveying teams actually measured part of this arc across the length of France. The idea of such a complicated and expensive procedure was that the unit would be defined in terms of something everlasting, instead of something temporary, like the King's foot. During the nineteenth century, an international team of scientists created a number of platinum–iridium bars, which are still preserved in temperature-controlled vaults. The meter was defined to be the distance between two scribe marks on one of these bars. Before the nineteenth century was over, the distance had been measured in terms of the number of wavelengths of a particular color of light. Since 1960, the internationally accepted meter is defined to be 1,650,763.73 wavelengths of the orange–red line in the spectrum of krypton-86. Such a standard is not subject to theft or wear and tear, and the optical measuring devices used for comparison are fairly inexpensive to produce and use.

The meter is subdivided in the metric method, by factors of 10. The prefixes used for names smaller than 1 are Latin; the prefixes used for larger multiples than 1 are Greek. Thus, 1/100 of a meter is a centimeter (cm). The symbol for meter itself is simply m. 1/10 of a meter is a decimeter (dm), but this term is rarely used in science or industry. 1/1000 of a meter is a millimeter (mm). There is a jump to the next commonly used subdivision and that is to the millionth of a meter, the micrometer ($\mu$m), or micron ($\mu$). The only commonly used multiple for a distance longer than 1 m is for 1000 m, the kilometer (km). A kilometer is about $\frac{5}{8}$ of a mile.

Measure yourself in terms of the meter and its subunits. In the list that follows there are two answer spaces. The first is for a measurement that you should make using just your eyes—in other words, an estimate. Like all measurements, you answer will be incomplete without a statement of the errors. Remember, when you give your plus or minus uncertainty limits, you are guaranteeing that a more careful measurement will be within those limits. After you have made your estimates, use a meter stick and any auxiliary instruments that you need to measure the same quantities. Once again, include the error limits, which, this time, should be much smaller than for the estimates.

| Measurement of Your: | Estimate | With Meter Stick or Other Instruments |
|---|---|---|
| Height | | |
| Circumference at waist | | |
| Foot length | | |
| Index finger width | | |
| Index nail thickness | | |
| Distance between tip of thumb and tip of the little finger when hand is fully extended (the span) | | |
| Elbow to wrist when hand is at right angles to arm | | |
| Thickness of one page of this book | | |
| Distance between pupils of your eyes | | |
| Distance between your nose and your fingertip when your arm is held out to the side | | |
| Your pace (the double-step) | | |

Several of these measurements require special techniques. To measure your height, for instance, it's a good idea to stand up straight against a wall and have a friend make a mark on the wall at the level of the top of your head. To measure your circumference, you will need something besides a rigid meter stick. Try a string, a belt, or your ingenuity. Convert the distance between your elbow and your wrist into inches and feet. There are 2.54 cm in 1 in. You probably will not be able to measure precisely the thickness of this page with a centimeter ruler. Can you, however, measure precisely the thickness of 100 pages? When measuring your pace, do not simply stride two steps and then measure that distance. The outdoorsman's pace is made up of normal walking steps of the type that you could keep up during a long walk. To measure the distance, it is best to measure 10 paces, and then divide by 10. To measure the distance between the pupils of your eyes, you will need either a companion or a mirror. However, if you use a mirror, and lay the centimeter ruler down on the mirror to measure the distance between the images of your pupils, your answer will be exactly half the actual distance.

## ABSOLUTE AND PERCENTAGE ERRORS

Errors are not mistakes; they are limits on the uncertainties of measurement. Through experience based on the type of instrument that you use, your skill in using it, and your need for precision, you determine that a measurement has a particular value plus or minus a certain error. For instance, your height may be $1.84 \pm 0.01$ m. You are guaranteeing that every measurement of your height that might be made, whether by somebody else or by using some other instrument, would be within 1 cm of 184 cm. This spread of uncertainty is called the *absolute error*. It is an uncertainty based on pessimism. You may very well feel that your original measurement was really good to $\pm 0.2$ mm, but for one reason or another, you can't guarantee it better than to $\pm 1$ cm. In sophisticated error analysis, there are other types of errors. The standard deviation, for example, is an uncertainty such that $\frac{2}{3}$ of all subsequent measurements might be expected to fall within the region of plus or minus a standard deviation. Since we expect all measured values to fall within plus or minus the absolute error, the absolute error is larger than the standard deviation. Furthermore, the standard deviation can be used appropriately only under restricted conditions where deviations in the data are determined by chance.

When you were measuring the lengths of parts of your body, suppose you had an absolute error of $\pm 1$ cm. Is that a large error? Were you being sloppy? If you were measuring the width of your little fingernail, then that error is quite gross. You would have had something like $1 \pm 1$ cm. On the other hand, if that absolute error is connected with the measurement of your height, then the measurement is much more precise. $184 \pm 1$ cm means that you have measured your height to about 1 part in 200. The absolute error is the same in both cases, but the relative error is much less in the latter case.

The *relative* error, or *fractional* error, or *percentage* error, is a comparison of the absolute error with the value of the measurement itself. In the case of measuring your fingernail, the relative error would be 1 part out of 1; the fractional error would be 1; the percentage error would be 100%. In the case of measuring your height, the relative error is 1 part out of 184; the fractional error is 1/184; the percentage error is about $\frac{1}{2}$%.

Should you be worried if there is a gram of mercury salt in your town's drinking water? That obviously depends not just on the quantity of mercury, but on its dilution. Is the pollution 1 part in 1000, 1 part in 1,000,000, or 1 part in 1,000,000,000? The relative concentration is the crucial factor.

Let's give some general symbols to these ideas. Suppose that you measure the length of something and determine that it has a length of $L$ (m, cm, or some special units). You judge that the uncertainty in your measurement is $\Delta L$. Your measurement, then, is that the length is equal to $L \pm \Delta L$. The absolute error is $\Delta L$. The fractional error is $\Delta L/L$. The percentage error is $(\Delta L/L) \times 100\%$. If the fractional error is $\frac{1}{10}$, then the percentage error is 10%.

Here are some exercises in taking percentages. If you already know how to take percentages, it will take you only a couple of minutes to find the answers. If you have trouble with any of them, then study our answers until you are sure that you know the process.

## Fractions, Percentages, and Parts Per ...

Percentage means parts per hundred:

$$10\% = \tfrac{1}{10} = 0.1 = 10 \text{ parts per hundred}$$

To find percentage, multiply a fraction or a decimal by 100:

$$\tfrac{1}{4} = \tfrac{1}{4} 100\% = 25\% = 25 \text{ parts per hundred}$$

$$0.4 = 0.4 \times 100\% = 40\% = 40 \text{ parts per hundred}$$

Change the following fractions and decimals to percent:

0.13 =      (13%)

$\tfrac{1}{8}$ =      (13%)

0.06 =      (6%)

$\tfrac{1}{6}$ =      (17%)

$1 \times 10^{-3}$ =      (0.1%)

To change percentage to a decimal, divide by 100. Change the following percentages to decimals:

38% = [ ] (0.38)

0.03% = [ ] $(3 \times 10^{-4})$

6.8% = [ ] (0.068)

If the maximum allowable concentration of a toxin in drinking water is two parts per million, what is the percentage maximum? Divide both sides by 10,000:

$$\frac{2}{10,000} \text{ parts per } \frac{\text{million}}{10,000} = 2 \times 10^{-4} \text{ parts per } 100 = 2 \times 10^{-4}\%$$

If the concentration is three parts per thousand, what is the percentage concentration?

[ ] (0.3%)

If the percentage concentration is $4 \times 10^{-3}\%$, how many parts per million are there?

[ ] (40)

If speed increases from 20 to 25 m/s, what is the percentage increase?

$$\tfrac{5}{20} = 5 \text{ parts per } 20 = \tfrac{5}{20} 100\% = 25\% \text{ increase}$$

If the speed then decreases from 25 to 20 m/s, what is the percentage decrease?

$$\tfrac{-5}{25} = -5 \text{ parts per } 25 = \tfrac{-5}{25} 100\% = 20\% \text{ decrease}$$

What is the percentage profit if a person buys an object for $8 and sells it for $10?

[ ] (25%)

What is the percentage loss if a person sells an object for $8 that he bought for $10?

[ ] (20%)

If the tax rate rose by 10% in 1979 but fell by 10% in 1980, do you win or lose? [ ]

If the original tax rate was $r\%$, the 1979 tax rate would be $r(1 + 0.1)\%$. In 1980 this new rate would fall to $[r(1 + 0.1)](1 - 0.1)\% = r(1 - 0.01)\% = 0.99r\%$. You win.

If the tax rate first falls by 10%, then rises by 10%, do you win or lose? [ ]

## ERRORS WHEN MEASURED VALUES ARE COMBINED

So far we have been talking about taking a single measurement—a length or a mass or a time. What happens if you have to combine measurements? For instance, suppose that you want to use a meter stick to measure the length of a table that is 137 cm long. You have to line up one end of the meter stick with the table, making sure that there is no parallax in the way that you are viewing the end, and also taking into account the fact that the edge of the table may be rounded. If it is, then part of your absolute error must be attributed to the thing being measured rather than to the measurer. At the other end of the meter stick, you must make a mark of some kind. The thickness of this mark and the care with which you make it will also contribute to the absolute error of your first measurement. Now you must move the meter stick and make a second measurement. You line up the first end of the meter stick with the mark that you made on the table, and then you must line up the other end of the table with the meter stick, trying to avoid parallax again. There is an absolute error connected with both the first measurement and the second one. How do you combine these two errors? If you decide that you could have been off by $\pm 2$ mm in each case, then the safe, the pessimistic, assumption must be that you could have been off by as much as $\pm 4$ mm for the combination. In other words, *in the addition of two measured quantities (or in subtraction) you must add the absolute errors.* You may think that this method is too pessimistic. If your first measurement was too large by 2 mm, perhaps your second measurement is small by 2 mm, thus canceling out the error. However you can't count on this cancellation. The safe rule is to add the absolute errors. If you are subtracting two measurements, this rule can lead to very large relative errors. Suppose that you estimate the length of a piece of cloth to be $10 \pm 1$ cm. If you now tear off a piece which you estimate to have a length of $7 \pm 1$ cm, then the best that you can say about the remaining piece is that it has a length of $3 \pm 2$ cm. The fractional error of your final information is $\frac{2}{3}$. Such a situation is not unusual in real technical measurements. Subtraction of two large numbers commonly leads to an answer with a very large relative error.

What happens to errors if you multiply measurements together? Suppose that you measure the length and width of a table and want to know the area. First, you have to make some assumption about the shape of the table. If you assume that it is rectangular, or make several measurements to make sure, then the area is simply the product of length and width:

$$A = LW$$

However there are errors attached to each of those two measurements. The area should be written

$A = (L + l)(W + w) = LW + lW + wL + lw$. Note how these algebraic terms correspond to the geometric areas of the diagram below. Even though the absolute errors $l$ and $w$ are small, their contribution to the absolute error in the area can be large. Let's find the fractional error in the area:

$$\frac{\text{error in area}}{\text{area}} = \frac{lW + wL + lw}{LW}$$

If we divide the numerator by the denominator, we get

$$\frac{l}{L} + \frac{w}{W} + \frac{lw}{LW}$$

If the fractional errors in length and width are very small, then their product is even smaller. Under these circumstances we can neglect the term $lw/LW$ in comparison with other fractions. Our rule for combining errors in multiplication is, therefore, *the fractional error in the product is equal to the sum of the fractional errors in the individual measurements.* If we multiply all these fractions by 100, we can specify the rule in terms of percentages: the percentage error in the product is the sum of the percentage errors of the individual measurements. The same rule applies to division as well as to multiplication.

Once again, this rule is one of maximum pessimism. It assumes that if the measured length was too long, so is the measured width, whereas in practice there may very well be cancellations. Nevertheless, the rule gives the safe way to specify the maximum uncertainty.

Let's take a specific example with the rectangular area shown in Fig. 3-2. Its length is equal to $10 \pm 1$ m and its width is equal to $5 \pm 1$ m. The absolute errors in both cases are $\pm 1$ m. The percentage error in the length is $\pm 10\%$ and in the width, $\pm 20\%$. The measured area, therefore, is $50$ m$^2 \pm 30\%$. The percentage errors have been added to find the percentage error of the product. The absolute error in the area is $\pm 30\%$ of $50$ m$^2$, or $\pm 15$ m$^2$.

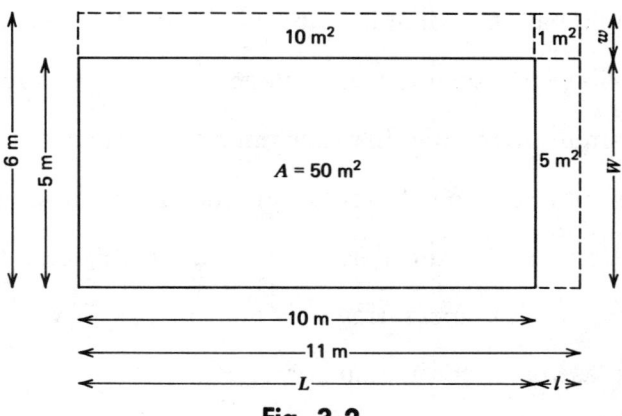

**Fig. 3-2**

Let's check to make sure that our method has given a reasonable answer. According to our original uncertainties, the length could have been as large as 11 m and the width as large as 6 m. Therefore, the area could have been as large as 66 m². On the other hand, the length could have been as small as 9 m and the width as small as 4 m. Therefore, the area could have been as small as 36 m². These limits agree well with those given by our method of adding absolute errors. Why don't the two methods agree exactly? What was ignored in our original derivation?

## EXERCISES IN DETERMINING ERRORS OF MULTIPLIED QUANTITIES

1. Use a centimeter ruler to measure the area of the cover of this book. Note that the absolute errors in length and width are probably determined more by the shape of the edge of the book than by your measuring instrument.

2. Measure the floor area of a room (any room that's convenient) by pacing. There will probably be some error associated with your uncertainty as to the fraction of paces when you reach a wall. There is probably also an error associated with your knowledge of the conversion constant that turns the number of your paces into meters. You must use your own judgment in reconciling these two types of errors to obtain reasonable errors for the measured length and width. Probably your errors for the measurement of the room will be much larger than for the measurement of the size of the book cover.

3. Get a ball of some kind (tennis, golf, baseball, and so forth). Measure the diameter by lining up the ball with a meter stick. Don't use calipers. The volume of a sphere is $\frac{4}{3}\pi r^3$. You may want to change that formula into one using diameter. If you use either formula for the volume of the ball, you are assuming that a geometric sphere is a good model for the ball you have chosen. It probably is, although note that a golf ball has dimples, and a tennis ball has fuzz. If you determine the percentage error in the diameter, then the percentage error in the volume is three times as large. That is because you are multiplying the diameter times the diameter times the diameter. Our rule for combining errors in multiplication calls for adding the individual percentage errors. In this case, it's the same percentage error. It is also particularly true in this case that our rule of maximum pessimism is justified. Since you are measuring only one variable, if your measurement is too high it is automatically too high as you cube the number.

4. Measure the volume of a sharpened pencil or a used piece of chalk. Remember that by itself, this direction is meaningless. How much precision should be obtained? For what purpose is the information needed? If you do not know the answer to those questions, you do not know how to proceed. One way to specify the precision required is to specify the money available to get it. In this case, we will specify time, which is usually equivalent to money. Find the volume of the chalk or pencil using no more than 5 min for the measurement. Under such circumstances, you will probably want to assume a model of a cylinder for the shape of the pencil or chalk. The model probably will not be very good. On the other hand, you have only 5 min. The formula for the volume of a cylinder is $\pi r^2 h$. Because of the poverty of your model, you are certainly not justified in using anything except a ruler to measure the diameter and the height. You may want to change the formula so that it is based on diameter. Incidentally, if you have a 10% error in the diameter, then you also have a 10% error in the radius. If you assign an absolute error of 2 mm to the diameter, then your absolute error in the radius is 1 mm. The relationship between radius and diameter is purely a mathematical one and does not follow the rules of experimental data combination. Remember that in calculating the volume you are multiplying the diameter times the diameter times the height; therefore, you must double the percentage error of the diameter and add that to the percentage error of the height. Before making the calculation of the volume, take a look at the pencil and compare it with the size of a cubic centimeter, shown actual size in Fig. 3-3. Approximately how many cubic centimeters do you expect the pencil or chalk to be? If your calculation is wildly different, then you had better check your method or your estimate.

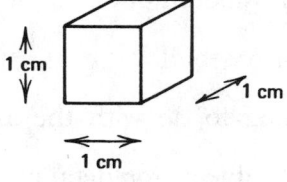

**Fig. 3-3**

# CHAPTER 4
# UNITS AND MEASUREMENT OF MASS

Mass and weight are two different things. Weight is the gravitational force exerted between two different objects, usually between an object, such as yourself, and the earth. Note, however, that weight is a force; mass is not. Mass has two different properties, seemingly quite different, yet on a profound level, the same thing. First, mass is the "charge" of the gravitational attraction, the property of matter that determines the magnitude of the attractive gravitational force between objects, e.g., between the sun and the earth. The unit of mass is the kilogram (kg). A kilogram *weighs* 2.2 lb. If you pick up a 5 lb bag of sugar at the store, it has a mass of a little over 2 kg. What is your mass in kilograms? Divide your weight in pounds by 2.2.

The other property of mass is called inertia. Objects resist having their velocities altered. The more massive the object, the more the resistance. At first glance, this inertial property seems to have nothing to do with weight. You can have a large door on a bank vault well-suspended on ball bearing hinges so that its weight does not affect its movement. Nevertheless, to get the door moving or to stop it once it is in motion requires a considerable amount of force. On a more familiar scale, it is clearly a lot harder to accelerate a shotput than it is a baseball.

The mass of an object has something to do with the amount of material in the object. The relationship and the exact definition are subjects for detailed study in a regular physics course. For now, we want to measure the mass of a number of familiar objects, so that we can get acquainted with the kilogram unit. Note, incidentally, that the basic unit of mass already has a multiplier in its name—kilo. You might think that the basic unit of mass would therefore be the gram, but international custom has decreed otherwise. The only multiples of the name, gram, that are commonly used are kilogram, gram, milligram, and microgram: kg, g, mg, $\mu$g. 1000 kg is a metric ton. Such a mass has a weight very close to the English ton, since 1000 kg would weigh 2200 lb.

Using pan scales, laundry scales, baby scales, or conversion units from weights that you know in pounds, find the mass of the following objects:

1. This book ☐ ± ☐

2. A pencil ☐ ± ☐

3. A nickel ☐ ± ☐

4. Your shoe ☐ ± ☐

5. A cup of water (not the cup—just the water) ☐ ± ☐

6. A quart of water ☐ ± ☐

7. A liter of water ☐ ± ☐

8. Your car ☐ ± ☐

9. A baseball ☐ ± ☐

10. A ping-pong ball ☐ ± ☐

## MASS DENSITY

Which is more massive, lead or styrofoam? That depends, of course, on how much lead and how much styrofoam. A toy lead soldier is less massive than a cubic meter of styrofoam. In comparing the "massiveness" of materials, we are usually concerned with comparing their "mass densities."

> The mass density is the mass per unit volume.

For SI units, density is expressed in kilograms per cubic meter ($kg/m^3$). In many cases, it is convenient to talk about mass density expressed in earlier metric units such as grams per cubic centimeter ($g/cm^3$).

It's very useful to have a feel for the density of common materials. Water has a density of 1 $g/cm^3$. Indeed, the gram was originally defined in terms of the mass of a cubic centimeter of water at its maximum density which occurs at a temperature of 4°C.

> A liter is equal to 1000 cm$^3$.

If the volume were a cube, it would have sides of 10 cm by 10 cm by 10 cm. The liter is approximately one quart. What is the mass of a liter of water? [____]. Remember, when it comes to liquids that are mostly water, "a pint's a pound the world around," and "a liter's a kilogram in every land."

Now let's express the density of water in SI units. What's the mass of a cubic meter of water? A cubic meter is a very large volume by human standards. Get a meter stick and roughly map out one cubic meter in one corner of your room. Note that if you filled that volume with water, it would be more than you usually have in your bathtub. How many cubic centimeters in one cubic meter? There would be 100 cm along each edge of the cube. Therefore, the cubic meter is composed of $100 \times 100 \times 100$ cm$^3$ = $1 \times 10^6$ cm$^3$. The mass of that much water is $1 \times 10^6$ g = $1 \times 10^3$ kg. 1000 kg weighs 2200 lb, a little over a ton. In SI units, the density of water is 1000 kg/m$^3$. In the table on p. 29 we give the densities of other common materials. Note, particularly, the values for the gases. A cubic meter of air has a considerable mass. Also note that lead is relatively un-dense. According to its place in the Periodic Table of the elements, it should have a density more like that of gold. The lead atoms are relatively large for their mass, and, therefore, their density is smaller than might be expected. Lead will float in mercury.

What's your density? You have already calculated your mass in kilograms by converting from your weight in pounds: your mass [____] ± [____] kg. It's harder to determine your volume than your mass. Here are some suggestions for measuring your volume. Judge what sort of cylinder you could just fit into. Assume a diameter and a height, and then calculate the volume of this cylinder:

diameter [____] ± [____] m    height [____] ± [____] m

volume = [____]

Don't forget that as you calculate the volume of your cylindrical capsule you must account for your estimated errors. The percentage errors for diameter and height must be properly combined. To calculate your density, divide your mass by your computed volume, once again taking into account the percentage errors in each:

density = [____] ± [____] kg/m$^3$

Whenever you compute a quantity you should check to make sure that it makes sense in terms of other things you know. For instance, does your computed density agree with the fact that you can just about float in water (perhaps you *can* float)? If your density is greater than that of water, you sink; if your density is less than that of water, you float. (Can you change your density?) Therefore, your estimated density, within your error margins, should be the same as that of water. If not, you must have misjudged the size of your capsule.

There is another easy way to measure your volume. If you have access to a bathtub, fill it with enough water so that you can completely submerge in it. When you are under, have a friendly collaborator mark the level of the water. After you get out, mark the lower level. When you dry off, you can compute your volume by multiplying the length, width, and height of the approximately rectangular cross section of the bathtub.

Check the density of copper given in the table that follows by measuring the density of copper pennies. If you have the proper instruments, you can measure the volume of an individual penny by measuring its thickness and diameter. You could measure its mass on a sensitive pan scale. Alternatively, you could take a large number of pennies, and find their total volume by submerging them in a measuring cup or a graduated beaker. Before they get wet, measure their mass on any convenient scale. Note that you can do the measurement in this second way with household devices. For instance, the scale could be a postage meter, a baby scale, or a laundry scale. The precision is poor, but then why did you want precision in the first place? Presumably, you merely wanted to find out how precisely you could measure the density of copper without spending more than 5 or 10 min on it.

| Material  | Density in kg/m$^3$ | Material | Density in kg/m$^3$ |
|-----------|---------------------|----------|---------------------|
| Air       | 1.25                | Mercury  | 13,500              |
| Hydrogen  | 0.09                | Iron     | 7900                |
| Helium    | 0.18                | Aluminum | 2700                |
| Water     | 1000                | Copper   | 8900                |
| Styrofoam | 100                 | Ice      | 920                 |
| Lead      | 11,300              | Gold     | 19,300              |

(At standard atmospheric pressure and at 0°C)

# CHAPTER 5
# THE SIMPLE MEASUREMENT OF TIME

Most of the SI units are human-sized. A meter is about as long as you could stretch with your arm. A kilogram is something massive enough to feel and yet easy to pick up. The unit of time is the second, a length of time very close to the period of a normal human heartbeat while at rest. You should check this for yourself. Grasp yourself by the throat, as shown in Fig. 5-1, with your index fingers pressing against your carotid arteries. (Doctors and nurses don't usually do it this way because it wouldn't look nice. However, the pulse is much stronger in your neck than it is in your wrist.)

Measure your heart rate while sitting still. The usual technique is to count the number of pulses in 30 s:

number of pulses in 30 s [ ]

number of pulses in 1 s [ ]

length of one heartbeat [ ] s

The standard symbol for seconds is simply s, not sec.

The smaller divisions of time are all decimals—millisecond (ms), microsecond ($\mu$s), nanosecond ($1 \times 10^{-9}$ s) (ns), and picosecond ($1 \times 10^{-12}$ s) (ps). For historical reasons, the multiples of time are usually those derived from the motion of the earth and are familiar to everyone. Note how strange it is, however, that the divisions are based on multiples of 12 rather than 10. There are 60 seconds in a minute, 60 minutes in an hour, and 24 hours in a day, all of these divisible by 12. It is true that there are not 360 days in a year, but there are 12 months in a year, and some cultures used to use the extra 5 days for holidays.

In a regular physics class, you will use many fancy instruments for measuring both short and long times with great precision. For many purposes, however, it is convenient to be able to measure short

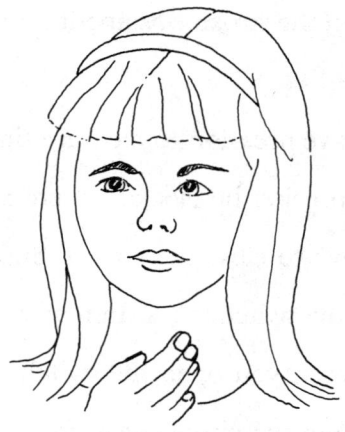

**Fig. 5-1**

times without any instrument. You could use your pulse, or you could learn to count seconds quite accurately by reciting "Mississippi-one, Mississippi-two, Mississippi-three," and so forth. Any four-syllable word will do, of course, If you don't like "Mississippi" try "locomotive" or "one-thousand-and-one, one-thousand-and-two," and so forth. Try this method while watching the sweep second hand of a clock. Test yourself by having a friend monitor clock times while you count out times of $\frac{1}{2}$ s, 3 s, 10 s, and 30 s:

$\frac{1}{2}$ s [ ] actual seconds

3 s [ ] actual seconds

10 s [ ] actual seconds

30 s [ ] actual seconds

As for the fractional seconds, note that with a total of five syllables, you are dividing 1 s into five parts, or 0.2 s. Try this method to find out how long it takes a solid object to drop 1 m, and then 2 m.

counted time for an object to fall 1 m [ ] s

counted time for an object to fall 2 m [ ] s

A well-known use of counting out time is to plot the approach of thunderstorms. The speed of sound in air is about 330 m/s, or about 1000 ft/s. Since you see the lightning flash almost instantaneously, you can measure the distance to the flash by counting down the time until the thunder arrives. If your count gets up to 5, then the stroke was about a mile away. If the count is under $\frac{1}{5}$ s, you probably should have been elsewhere.

In many laboratories, you will have occasion to measure time intervals by starting and stopping a clock with a switch. No matter how precise the clock is, there is bound to be some uncertainty about the time it takes your muscles to move to close the switch. Suppose that the time between seeing an event and the time your fingers close the switch is $\frac{1}{5}$ s. In many cases, this delay will be compensated at the other end of the timing period when you open the switch. But you can't count on it. The visual signal for starting may be different from the visual signal for stopping. Or you may be controlling the start with your muscles already tensed, and then have to anticipate when you will see the signal for stopping. In any case, you ought to measure the approximate time for your reflex actions. These will be different for head-to-fingers than they will for head-to-toes (an electrical signal is traveling along nerves, and the reaction time is roughly proportional to the distance traveled.)

Here's a time-tested method of measuring reflex times for your fingers when directed by your eyes. First, what do you think your reaction time is? Write down an estimate: ⬚ s. Have someone hold a meter stick vertically by a table as shown in Fig. 5-2. Place your thumb and index finger on either side of the stick so that you can grab it as soon as it starts to drop. Have the other person let go of the meter stick at random times in such a way that you can see the meter stick dropping but not the motion of the other person's hand. Measure the number of centimeters that the meter stick dropped before your finger and thumb stopped it. As you will learn in physics class, the relationship between the distance fallen and the time of fall for this case is

$$y = \tfrac{1}{2}gt^2$$

If the distance fallen $y$ is in centimeters, then $g = 980$ cm/s$^2$, and $t$ will be in seconds.

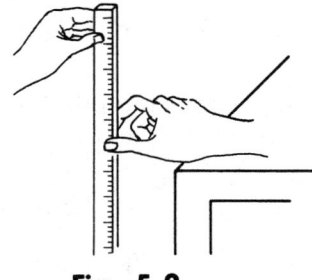

**Fig. 5-2**

| Distance Dropped | Time (in s) |
|---|---|
|  |  |
|  |  |
|  |  |
|  |  |
|  |  |
| Average reflex time |  |

In judging timing errors, you should remember the results of this experiment and assume that your error in such a measurement is comparable to the average that you calculated. How does your result compare with your initial estimate?

# CHAPTER 6
# UNITS AND DIMENSIONS

So far, we have talked about three main quantities: length, mass, and time. The SI units for these are meter, kilogram, and second, respectively. There are also some simple combinations of these quantities, which we have either mentioned or with which you are familiar. Surface area is found by multiplying a length times a length, yielding units, for instance, in square meters (m$^2$). Volume is measured in cubic meters (m$^3$). Density is mass per unit volume, measured in kilograms per cubic meter (kg/m$^3$). We have not yet defined speed or velocity, but it is a quantity used in everyday life with common units—miles per hour (mph), meters per second (m/s).

When talking about physical quantities, it is frequently useful to distinguish between their *dimensions* and their *units*. By "dimensionality" we mean whether the quantity is a length, a mass, a time, or some combination of those without regard to whether the measurement is made in meters or miles, kilograms or grams, seconds or years. A large number of quantities can be described in terms of the dimensions of length, mass, and time ($L$, $M$, $T$). There is nothing sacred about these three particular dimensions. It is possible, for instance, to choose velocity as a basic dimension, and it might even make more sense, since there is a natural velocity in the universe—the speed of light. Furthermore, in describing electrical quantities, it is necessary to bring in a fourth basic dimension, usually electric charge. In some types of analysis, it is also convenient to use temperature as an independent dimension. For our purposes, we can illustrate the principles by simply using $L$, $M$, and $T$. The dimension of area is then $L^2$; the dimension of mass density is $ML^{-3}$; the dimension of velocity is $LT^{-1}$.

Assigning dimensions to physical quantities may seem like an unnecessary exercise. However throughout the rest of this book we will demonstrate the usefulness of the technique. Whenever we develop an equation, we will first check to make sure that the dimensions on the left-hand side are the same as the dimensions on the right-hand side. Otherwise, we might end up equating apples to horses, or at least kilograms to velocities. Some examples of formulas to be developed later are given in the exercises. You can easily check their dimensionality without knowing the details of the formula. Just because the dimensions of a formula are correct does not guarantee that the other features are also correct. For instance, the formula for the area of a circle is $A = \pi r^2$. The dimensions of the left-hand side are $L^2$. Evidently, this is also true of the right-hand side, since there is a radius to be squared. Dimensional analysis cannot tell us, however, whether $\pi$ is the correct constant.

Dimensions have great generality, while units are very particular. In any technical work, it does no good at all to cite the area of something as $1.3 \times 10^4$. $1.3 \times 10^4$ *what?*—square meters? square miles? A number without units is meaningless.

Unfortunately, for technical reasons and historical reasons there are many units for each variable. In the Appendix at the back of this book, we provide tables for conversion of units. Even with these or similar tables, the conversion of units requires some routine techniques. The safest, most foolproof method of converting units is as follows. Always express the size of a quantity with its units following in fractional form. For instance, when you are traveling at the national speed limit, or within the margin of grace, your speed might be $60 \frac{\text{miles}}{\text{hour}}$. Suppose that you want to convert these units to meters per second. This particular conversion factor is listed in the Appendix at the back of the book, but let's do it here longhand. The trick is to multiply the fraction by other fractions *each of which is equal to 1*. In this case, since we want to eliminate miles in the numerator, let's convert the miles to feet:

$$60 \frac{\text{miles}}{\text{hour}} \times \left( \frac{5280 \text{ ft}}{1 \text{ mile}} \right)$$

Note that the fraction in parentheses does indeed have the value of 1. We do not change our original quantity by multiplying it by 1. At this point, if we carried out the multiplication, we would have the speed expressed in terms of feet per hour. The units of miles in the numerator and denominator have canceled. Now we want to get rid of the feet in the numerator and work our way toward having meters in the numerator. Once again, we could make the conversion in one jump, but let's take it slowly for the practice. Since there are 12 inches to a foot, we will multiply by the fraction $\left( \frac{12 \text{ in.}}{1 \text{ ft}} \right)$. If we multiplied by that fraction, we would have our speed in inches per hour. Let's keep working our way toward meters. We will multiply by the fraction $\left( \frac{2.54 \text{ cm}}{1 \text{ in.}} \right)$. One more conversion is necessary to put meters in the numerator. We multiply by the fraction $\left( \frac{1 \text{ m}}{100 \text{ cm}} \right)$. If we multiplied all those fractions together, we would have the speed in meters per hour. To get rid of hours in the denominator, we can multiply by the fraction $\left(\frac{1 \text{ h}}{3600 \text{ s}}\right)$. Now let's string all those fractions together.

$$60\frac{\text{miles}}{\text{h}} \times \left(\frac{5280 \text{ ft}}{1 \text{ mile}}\right) \times \left(\frac{12 \text{ in.}}{1 \text{ ft}}\right) \times \left(\frac{2.54 \text{ cm}}{1 \text{ in.}}\right) \times \left(\frac{1 \text{ m}}{100 \text{ cm}}\right) \times \left(\frac{1 \text{ h}}{3600 \text{ s}}\right)$$

The units can now be canceled just as if they were algebraic quantities. Note that all of them cancel except for meters in the numerator and seconds in the denominator. When we perform the arithmetic of multiplying the numbers together, we get

$$60\frac{\text{miles}}{\text{h}} = 26.8\frac{\text{m}}{\text{s}}$$

This mechanical method of changing units is foolproof providing that you don't make any fool mistakes. As usual with all problems, when you get a final answer, take a close look at the number and its units. Does it make sense? In this case, consider how fast you can walk in meters per second. Surely you can cover at least 1 m/s. Since your speed must be much less than 60 miles per hour, is our final answer reasonable?

Try the following conversions:

1. $7.5\frac{\text{g}}{\text{cm}^3}$ = ▭ = ▭ $\frac{\text{kg}}{\text{m}^3}$

2. $60\frac{\text{miles}}{\text{h}}$ = ▭ = ▭ $\frac{\text{ft}}{\text{s}}$

3. Change 1 light-year to meters. (A light-year is the distance light travels in one year.) The speed of light ($c$) is $3.0 \times 10^8$ m/s.

4. A standard physics experiment uses a ticker-tape running through a doorbell clapper to measure time. The doorbell clapper leaves dots on the paper that is running through it. Suppose that the doorbell is running at 14 times per second. When the ticker-tape was being dragged through by a heavy lead weight falling straight down, the acceleration was 5.0 cm/dot$^2$. What was the acceleration in m/s$^2$?

5. Change atmospheric pressure from 14.7 lb/in.$^2$ to newtons/m$^2$ (N/m$^2$). 1 m = 39.37 in. 1 lb = 4.4 N.

# CHAPTER 7
# SOME GENERAL COMMENTS ON GRAPHS

Scientific data are represented on graphs as much as they are represented as formulas. In order to do any kind of scientific work, you must be able to create graphs as well as to interpret them. There are some standard conventions about graphs—rules of the road that everyone uses. For example, in graphing motion, you should always plot time on the horizontal axis. If you don't, your graph may not be technically wrong, but it will be hard for anyone else to read.

Graphing should almost never be postponed until data-taking is all over. Instead, it is a tool to be used while the data are being collected. Look at the following table of experimental values:

| $t$(s) | $x$(m) |
|---|---|
| $0 \pm 0.2$ | $0 \pm 0.1$ |
| $1 \pm 0.2$ | $2 \pm 0.1$ |
| $2 \pm 0.2$ | $8 \pm 0.2$ |
| $3 \pm 0.2$ | $14 \pm 0.2$ |
| $4 \pm 0.2$ | $32 \pm 0.3$ |
| $5 \pm 0.2$ | $50 \pm 0.4$ |

If you recorded your data only this way, you probably would not notice that something peculiar had happened. Now look at the graph in Fig. 7-1.

You can see immediately that the smooth curve has a dip in it at one point. If you had taken the data one day, and then plotted the graph a week later before handing in your laboratory report, you would not know what to do with this data point that apparently is wrong. Or is it wrong? Perhaps that

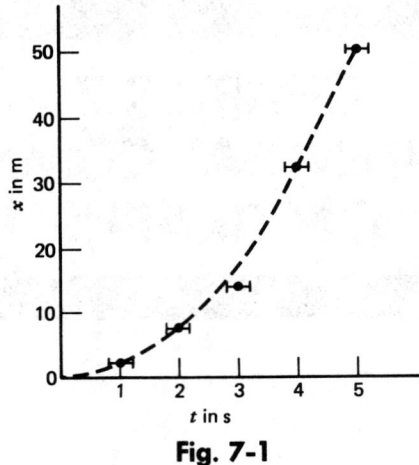

**Fig. 7-1**

unusual data point signifies a brand new phenomenon. If you throw it away, you lose the Nobel Prize. But how can you find out a week later? The apparatus has probably been torn down. Data should always be plotted point by point as the information comes in. This requirement doesn't mean that you must have good graph paper and produce a finished product. If you want to use good, expensive graph paper, you may, but in many cases it is perfectly suitable to sketch the axes, the divisions, and the graph itself. With a little bit of practice, you can make very useful sketches, whether or not you think you can draw a straight line.

There is another reason for sketching the graph of your results while you are taking the data. It can save you time and effort. Suppose that you are plotting the position of some moving object as a function of time. Should you measure its position every second, every ten seconds, every half hour? The answer depends on the nature of the motion. The motion, and so the answer, may change from the beginning to the end of the data-taking period. If you are plotting the data as you go, you can see whether there is a gap in the data that must be filled in, or whether you can safely assume that the motion is constant for certain periods of time. Figure 7-2 shows a graph with three data points.

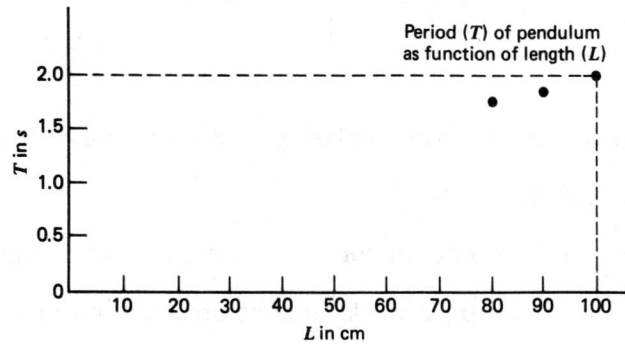

**Fig. 7-2**

As you can see, it probably would be a waste of time to make any more measurements in these same regions. The crucial point is, what is the behavior of the phenomenon for very short lengths? One piece of data, for instance, at 10 cm, is a guide to which measurements, if any, need to be taken next.

Here is a list of some of the definitions, conventions, and rules for graphs:

1. Every graph must have a title. Somewhere on a graph, there is room to describe what variable was being plotted as a function of what other variable. Don't be stingy on giving information. Four months later, just before final exams, you yourself may want to know which graph this was and what the conditions were.

2. The axes must be labeled with the names of the variables and the units being used. For instance, if you're plotting position as a function of time, then the vertical axis should be labeled $x$ (or $y$, or $z$), and the horizontal axis should be labeled $t$. Furthermore, the position axes must be labeled with m (for meters) or whatever other unit you are using, and the horizontal axis must be labeled s (for seconds) or whatever other unit you are using. An example of such labeling is shown in Fig. 7-3.

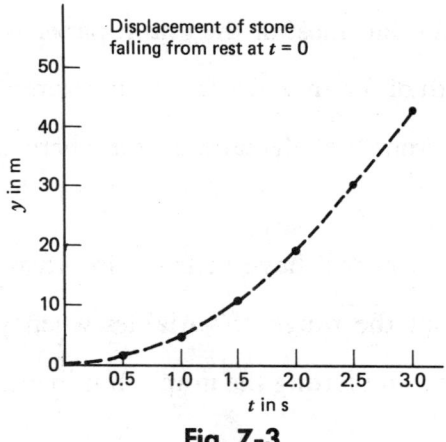

**Fig. 7-3**

3. The horizontal axis is usually used for the independent variable. This axis is called the *abscissa*. The dependent variable is usually plotted along the vertical axis, the *ordinate*. Which variable is dependent and which independent is often a matter of choice. If you are plotting the period of a pendulum as a function of its length, your usual procedure is to change the length of the pendulum string from 10 cm to 20 cm, and so forth. The length is then the independent variable. You measure whatever period is produced and so the period is dependent on the length. In this case, the length of the pendulum is plotted on the horizontal axis and the period along the vertical axis. In graphing

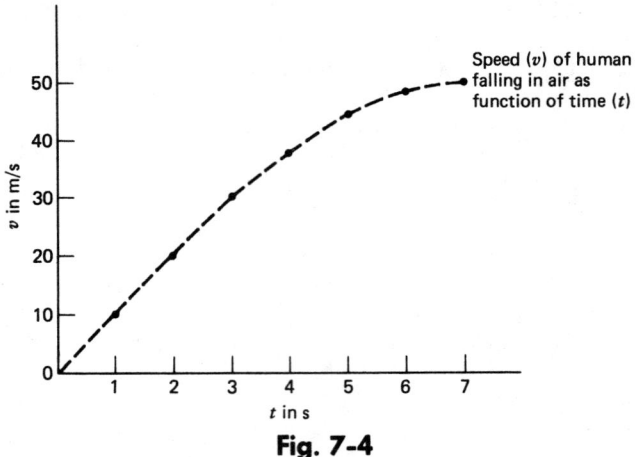

**Fig. 7-4**

motion, the convention, as we have already said, is to put time on the horizontal axis as if it were always the independent variable. Position, velocity, or acceleration would be plotted on the vertical axis as if they were dependent variables (see Fig. 7-4). This is the convention in spite of the fact that you might be measuring how long it takes a car to go from 0 to 10 mph, to 20 mph, and so forth. In such a case, the velocity is actually the independent variable and the time is dependent.

4. Choose the scales of your units so that most of the graph paper is used. The pair of graphs in Fig. 7-5 shows the same data, but with different scales for the horizontal axis. As you can see, it would be harder to interpret information from the left-hand graph where the curve is squeezed close to the axis.

In order to spread out a graph, you will have to know in advance the maximum and minimum values for the variables. Find out the range of variables when you first start taking data. Such information will be useful not only in plotting the graph, but in other aspects of the experiment also.

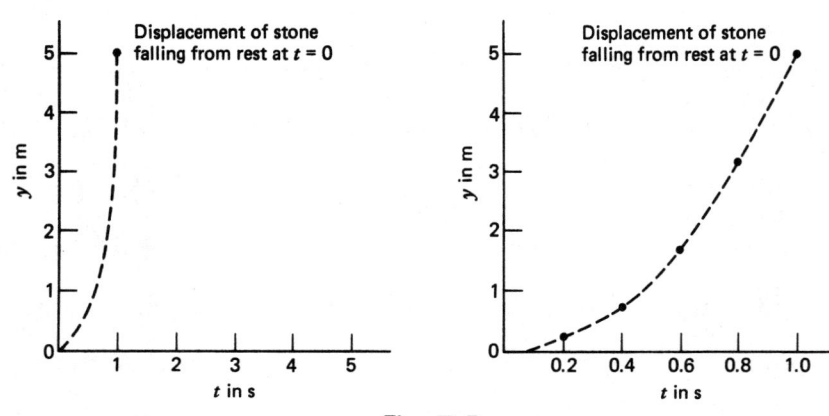

**Fig. 7-5**

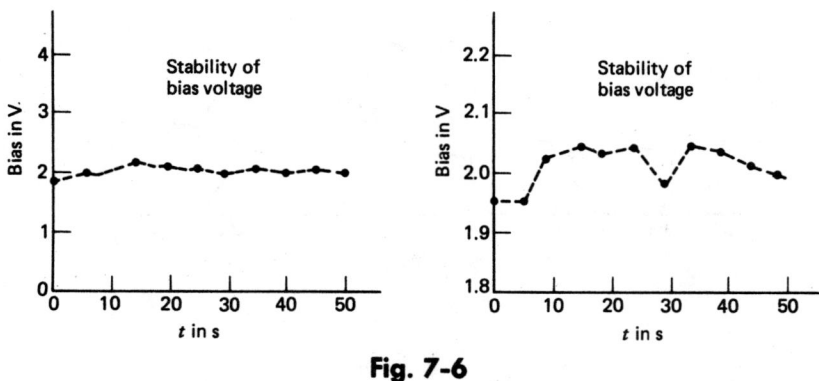

**Fig. 7-6**

For instance, you may want to make sure that some instrument will not have to go off scale or that the time needed is not longer than you have available.

5. There are lots of ways to make figures and graphs lie. Here are two examples involving the choice of scale. In the first pair of graphs (Fig. 7-6), the scale units are very different. With the large scale unit, it appears that the voltage of the system remains very constant. With the smaller unit, it is apparent that there are major fluctuations.

In the second pair of graphs (Fig. 7-7), the scale units are the same in both cases, but the zero point is different. In the left-hand graph, it appears that there are only small changes in the prices being shown. In the right-hand graph, it appears that the price changes are enormous. It is usually misleading to have a false zero on an axis of a graph. If you need to do so, you usually should draw attention to the fact in words.

6. Graphs are not just diagrams. In grade school you may have cut out a strip of paper equal to your height, written your name on it, and pasted it on the classroom wall. The array of paper strips shows something about the heights of the children in the classroom, but it is not a graph. A graph must show functional relationship between two or more variables. As one variable increases along

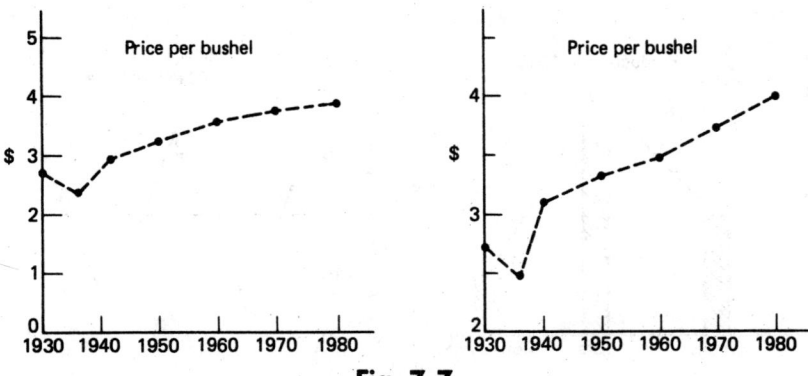

**Fig. 7-7**

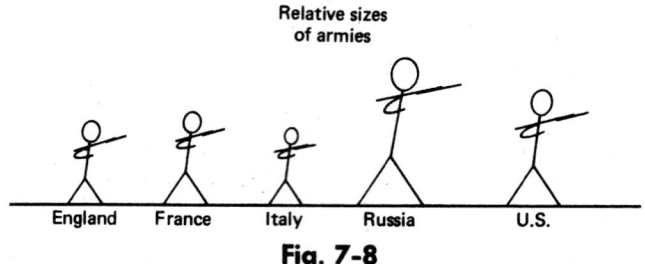

**Fig. 7-8**

one axis, the other variable changes according to some definite relationship. Sometimes newspapers have illustrations, particularly in the economic or political sections, that look something like graphs. For example, Fig. 7-8 shows the relative sizes of armies of several countries. There is no functional relationship between the data points, however. Indeed, such diagrams are usually very misleading. It is never clear whether the relative sizes are indicated by heights or the implied weights of the figures shown. (A soldier twice as tall must have a weight eight times as large.)

You can connect the data points on a graph with a curve only if functional dependence is assumed and if there is reason not to expect fine structure between points. Figure 7-9 shows two cases where data points cannot be simply connected. In the first, there is no functional dependence, and so the bar graph or histogram is not really a graph. In the second case, there is a functional dependence, but not the straight line (linear) type assumed.

7. Actually, there is no such thing as data "points" on a graph. As we have seen, there are usually errors in measurements. As you measure one variable at a particular value of another, there is uncertainty in both. Algebraically, these uncertainties are described in terms of plus or minus values (at $t = 3.0 \pm 0.2$ s, $x = 8.3 \pm 0.1$ m). The graph must faithfully show this uncertainty. The way this is done in all scientific and technical work, is to represent the uncertainties with horizontal and

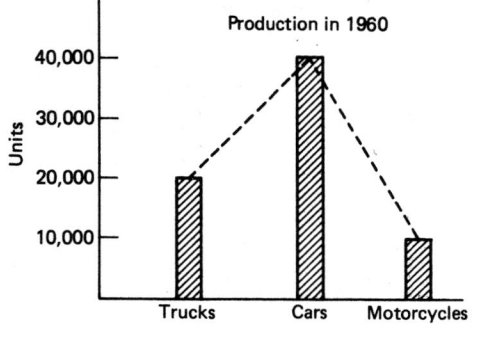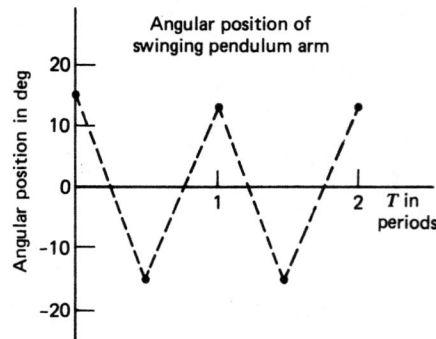

**Fig. 7-9**

vertical bars forming a data "region" instead of a data "point." An example of a table of values and the resulting graph is shown in Fig. 7-10.

| t | x |
|---|---|
| 0 | 0 |
| 1 ± 0.2 | 1 ± 0.1 |
| 2 ± 0.2 | 4 ± 0.4 |
| 3 ± 0.2 | 8 ± 0.8 |
| 4 ± 0.2 | 17 ± 2 |
| 5 ± 0.2 | 23 ± 2.5 |

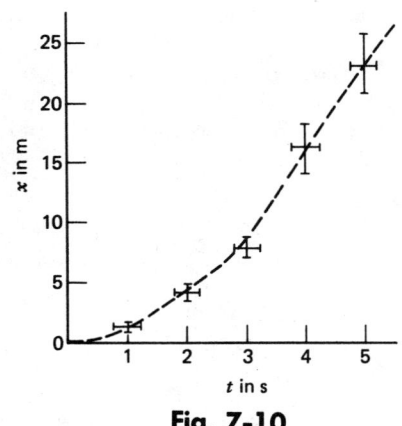

Fig. 7-10

When you first run across this method of representing errors graphically, there is great temptation simply to indicate the positions of the variables with crosses instead of with points or circles. That's not the point! With the plus and minus values of your data you are making a personal guarantee that any subsequent measurement would be found within the bounded region. In drawing a graph, you are making the same guarantee. In the real world, people count on those guarantees in designing airplanes, rockets, bridges, and other things that may fall on you. In drawing a curve connecting the data regions to illustrate functional dependence, you cannot expect that the curve will pass through the centers of the data crosses. If the curve does that, then you must have been kidding about the magnitudes of your errors. At best, you can hope that a smooth curve indicating functional dependence will not fall outside any of the data *regions*. Note the example in Fig. 7-11.

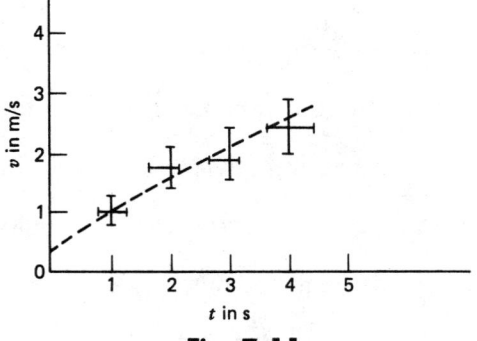

Fig. 7-11

## SOME GRAPHING EXERCISES

1. Here are some data showing the position as a function of time of an object. Graph the data and connect the data regions with a smooth curve. Remember to label the graph, label the axes, choose the correct scale so that most of the graph is used, and show the data bars correctly. First, however, take a look at the data given in the columnar form below. Can you tell if there is anything special about the data? After you plot the data, see if you can find anything special about it.

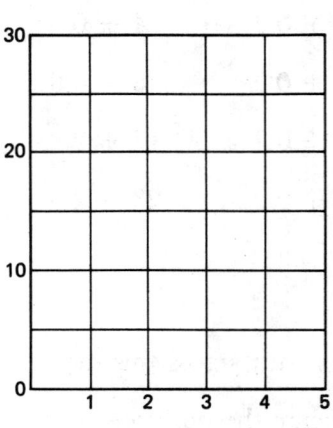

| $t(s)$ | $x(m)$ |
|---|---|
| $0 \pm 0.2$ | $0 \pm 0.1$ |
| $1 \pm 0.2$ | $2 \pm 0.1$ |
| $2 \pm 0.2$ | $5 \pm 0.2$ |
| $3 \pm 0.2$ | $15 \pm 0.2$ |
| $4 \pm 0.2$ | $20 \pm 0.3$ |
| $5 \pm 0.2$ | $30 \pm 0.4$ |

2. The two graphs below have different scales for the vertical axes. Draw on each graph the line representing the equation $x = \frac{1}{2}t$. Note how different the two graphs look, in spite of the fact that they represent the same information.

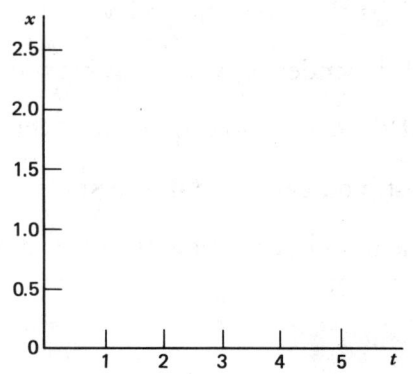
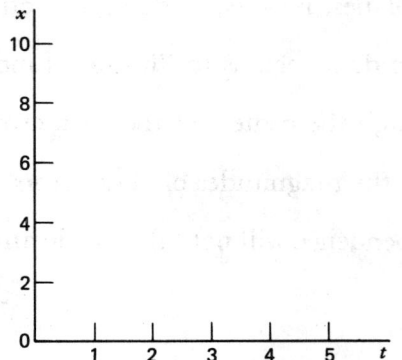

3. Get a measuring cup or a graduated cylinder and collect water from a faucet that has a slow but steady drip. Start timing the water collection when the first drop falls. Record the time as the water level reaches each subdivision of the measuring device. There will be errors involved; it will be hard to judge the exact position of the water level. Graph the data, making sure that you represent the error bars fairly. Draw a smooth line through the data regions. Can you tell from the appearance of the line whether the water was dripping at a steady rate?

# CHAPTER 8
# TRIGONOMETRY AND USEFUL ANGLES

The sinusoidal functions are functions of angles. An angle is a measure of the opening between two lines that cross. Frequently, we will use Greek letters to denote angles (see Fig. 8-1). This practice is standard in most physics texts. The most commonly used symbols are $\theta$ (theta—pronounced, thāta); $\phi$ (phi); $\alpha$ (alpha); $\beta$ (beta—pronounced, bāta); and $\gamma$ (gamma).

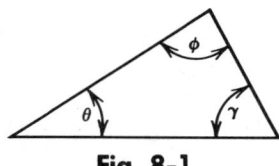

**Fig. 8-1**

For an operational (measurable) definition of an angle, we state that *the measure of an angle is the ratio of the arc it subtends to the radius arm*, as shown in Fig. 8-2. Note that this means that an angle has no dimensions. It is a length (of the arc) divided by a length (the radius out to the arc).

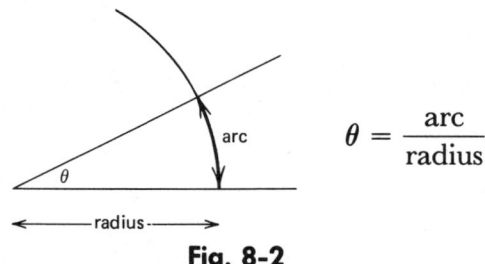

**Fig. 8-2**

Let's see what the *unit* angle would look like according to this definition. In Fig. 8-3, an arc length has been drawn equal to the radius. Take a string or piece of paper and measure for yourself that this is true for each of the concentric circles. As you can see, the definition works regardless of the size of the circle. Now, if $\theta = \dfrac{\text{arc}}{\text{radius}}$, and if in this case arc = radius, then in this case, $\theta = 1$. The name of this unit angle is the *radian*.

$$\text{arc}_1 = r_1 \quad \text{arc}_2 = r_2 \quad \text{arc}_3 = r_3$$

$$\theta = \dfrac{\text{arc}_1}{r_1} = \dfrac{\text{arc}_2}{r_2} = \dfrac{\text{arc}_3}{r_3} = 1$$

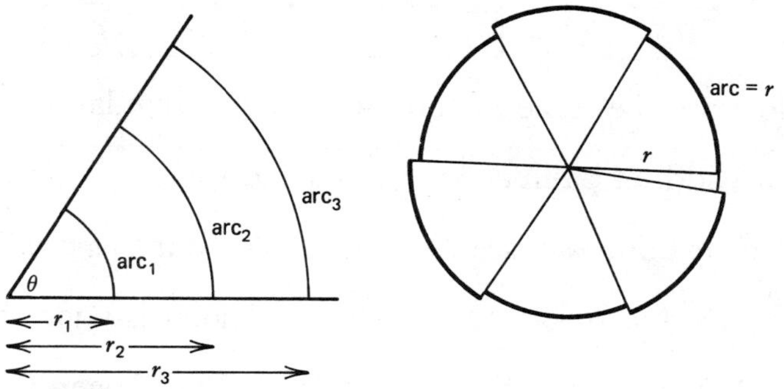

**Fig. 8-3**

**Question 8-1.*** We have sketched a number of unit angles in the circle in the diagram above. How many radians are there in a complete circle?

Long ago, the Babylonians (if not an earlier culture) divided circles into 360 equal divisions, or degrees. Why 360? Perhaps because 360 is divisible by a large number of factors—2, 3, 4, 5, 6, 8, 9, 10, 12, 15, and so forth. Consequently, it is easy to produce many different fractions of the circle in terms of degrees. Another reason for dividing a circle into 360 equal parts is that the earth revolves around the sun in 365 days. Of course, 365 is not equal to 360. On the other hand, with this round number, it is the case that the sun moves through the heavens (against the fixed stars) approximately 1° per day. Or, if you prefer a modern viewpoint, the earth moves around the sun about 1° per day.

If there are $2\pi$ radians (rad) in a circle, and also 360°, how many degrees per radian? ▭. This number is very useful to remember, at least to two significant figures. Also, note how close the number is to 60. $2\pi \approx 6$, so that there are about six radians in a circle. In technical work, we will measure angles in degrees and radians, and must be able to switch back and forth between them.

**Handling the Phenomena** It will be very convenient for you to become familiar with the approximate sizes of various angles. Get a protractor and, using a large scale on a large sheet of paper, draw the following angles: 90°, 80°, 1 rad, 45°, 30°, 20°, 10°, 5°, 1°.

By using the radian definition of an angle, it is easy to measure or estimate the angular width of an object. For instance, what is the angular width of your index fingernail when it is held at arm's

---

*Our answers or comments on questions are at the end of the chapter.

length from your eye? To find out, measure the actual width of your fingernail. ⬚ cm. What is the distance from your eye to your fingernail when your arm is extended with the finger up? ⬚ cm. Your fingernail subtends an angle whose size in radians is given by arc/radius. In this case, $\theta$ = ⬚ = ⬚ rad = ⬚ deg.

Now use the same method to find the angular width of your palm when held at arm's length. Angular width of palm = ⬚ rad = ⬚ deg.

The next time you see the moon, hold your fingernail up at arm's length and you will observe that it can cover the moon. Obviously, the moon is wider than your fingernail, but its *angular width* is smaller. Find out if the fingernail on your little finger, when held at arm's length, can just obscure the moon. If you become familiar with that fact, try obscuring the giant harvest moon when it is close to the horizon. The apparent huge size of the harvest moon is an optical illusion. You can break that illusion for yourself by measuring that the angular width of the moon has not really changed.

## THE TRIG FUNCTIONS

We have spent considerable time defining what we mean by angle. Now let's define what we mean by the sine, cosine, and tangent of an angle. There are several ways to do this mathematically, but we will use the definition based on the properties of a right triangle (a right triangle has one 90° angle). An angle of less than 90° is called an *acute angle*.

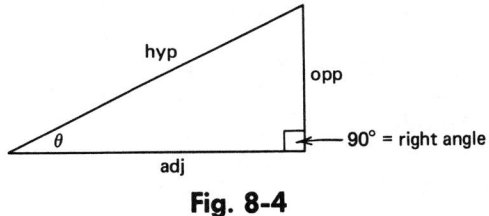

**Fig. 8-4**

In Fig. 8-4 there is a labeled right triangle. The *hypotenuse* is the name of the side opposite the 90° angle. We will define the functions in terms of angle $\theta$. The sides opposite (opp) and adjacent (adj) to $\theta$ are labeled. The sine, cosine, and tangent of $\theta$ are defined in terms of ratios of the sides. Note the standard abbreviations for the sinusoidal functions (sin is pronounced sine and cos is pronounced cosine).

$$\sin \theta = \frac{\text{opp}}{\text{hyp}} \qquad \cos \theta = \frac{\text{adj}}{\text{hyp}} \qquad \tan \theta = \frac{\text{opp}}{\text{adj}}$$

**Handling the Phenomena** Use a protractor to draw large *right* triangles containing the following angles: $\theta = 10°, 30°, 45°, 60°$, and $80°$. With a ruler, actually measure the ratio of lengths to find the value of $\sin \theta$, $\cos \theta$, and $\tan \theta$ for each of these triangles. (Before you draw the triangles, consider carefully how many you have to draw in order to obtain the required angles.) From your measurements, fill out the following table of values. Note that values are also required for $\theta = 0°$ and $90°$, even though you have not drawn triangles containing those angles.

**Measured Values**

| $\theta$ | $\sin \theta$ | $\cos \theta$ | $\tan \theta$ |
|---|---|---|---|
| 0 | | | |
| 10 | | | |
| 30 | | | |
| 45 | | | |
| 60 | | | |
| 80 | | | |
| 90 | | | |

There are some easy ways to calculate analytically the sines and cosines of certain angles without going to the trouble of actually measuring them. Look at the following diagram of an *equilateral* triangle. All three sides are equal and therefore each of the angles of the triangle is equal to 60°. If you drop a perpendicular from the top vertex, you bisect the triangle into two other triangles, each of them now a 60–30–90 triangle. If the sides of the equilateral triangle originally were two units in length, then the base of each 30–60–90 triangle is one unit in length. The length of the perpendicular bisector can be found using the Pythagorean theorem. For a right triangle, $c^2 = a^2 + b^2$, as shown in the diagram. The derivation in Fig. 8-5 shows that the length of the perpendicular bisector is 1.7 units. Therefore, $\sin 30° = \frac{1}{2}$ and $\sin 60° = 1.7/2 = 0.87$.

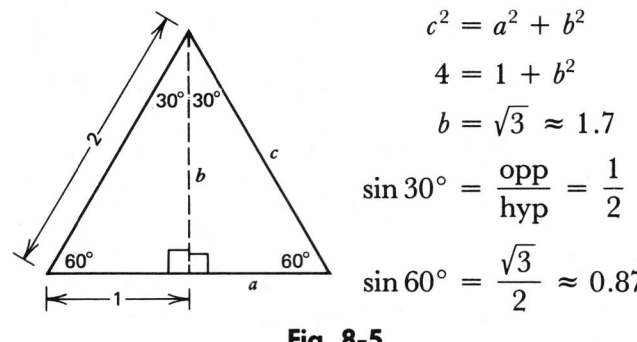

$$c^2 = a^2 + b^2$$
$$4 = 1 + b^2$$
$$b = \sqrt{3} \approx 1.7$$
$$\sin 30° = \frac{\text{opp}}{\text{hyp}} = \frac{1}{2}$$
$$\sin 60° = \frac{\sqrt{3}}{2} \approx 0.87$$

**Fig. 8-5**

Another geometrical construction can give us the sine and cosine of 45°. The triangle shown in Fig. 8-6 is *isosceles*, with two sides the same length. Let each side have unit length 1. The angle between them is 90°. Use the Pythagorean theorem to find the length of the hypotenuse. Hypotenuse = _____.

$\sin 45° = \cos 45°$. Why? _____

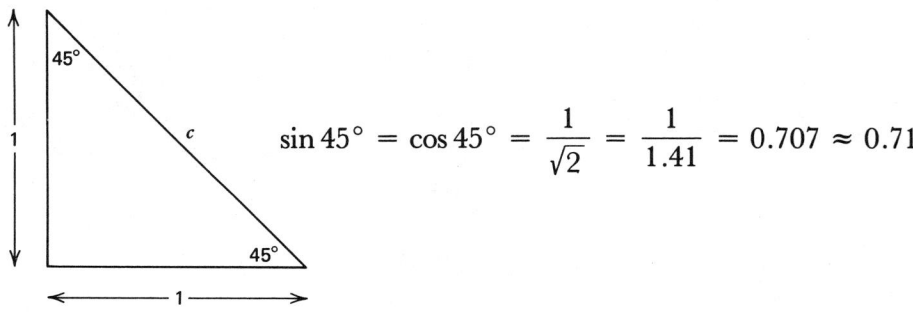

$$\sin 45° = \cos 45° = \frac{1}{\sqrt{2}} = \frac{1}{1.41} = 0.707 \approx 0.71$$

**Fig. 8-6**

Now you have enough information, both from measurements of triangles and from these analytical derivations, to plot the graphs of $\sin \theta$ versus $\theta$, $\cos \theta$ versus $\theta$, and $\tan \theta$ versus $\theta$. Record these data on the following graphs for the angles between 0° and 90°:

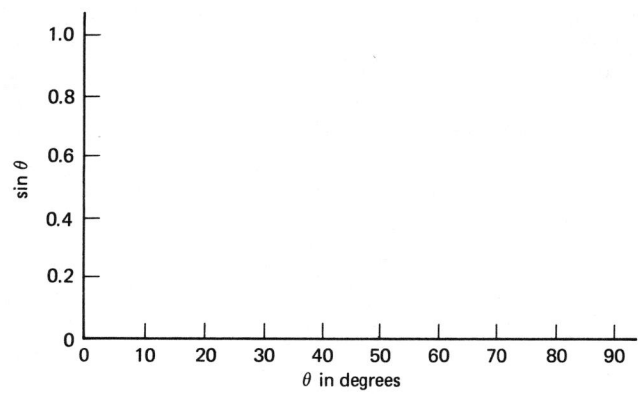

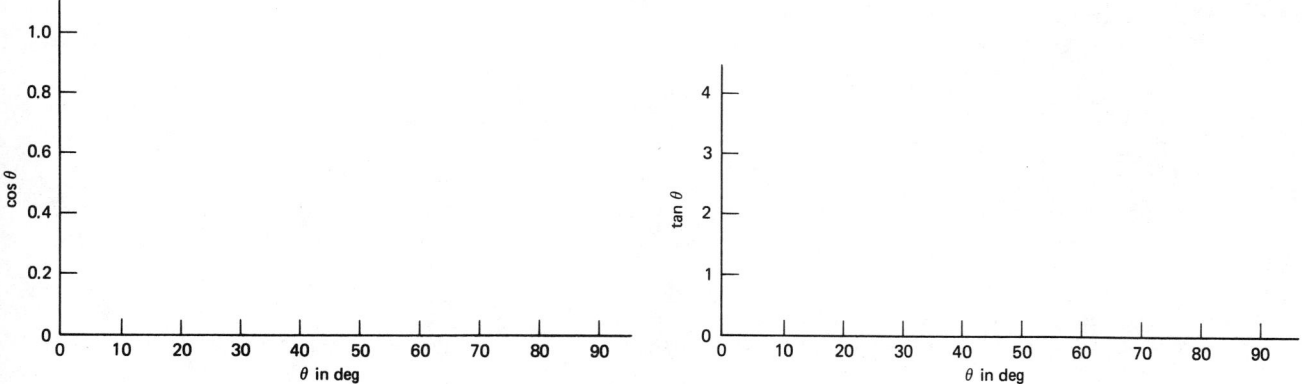

Incidentally, note the way that $\tan\theta$ takes on values larger than 1. Another way to define $\tan\theta$ is

$$\tan\theta = \frac{\text{opp}}{\text{adj}} = \frac{\text{opp/hyp}}{\text{adj/hyp}} = \frac{\sin\theta}{\cos\theta}$$

When $\theta$ is small, $\sin\theta$ is small and $\cos\theta$ is nearly 1. Therefore, $\tan\theta$ is close to 0. However, when $\theta = 45°$, $\sin 45° = \cos 45°$, and $\tan 45° = 1$. From that point on, as $\theta$ increases toward 90°, $\sin\theta$ gets closer to 1, but $\cos\theta$ goes to 0. Consequently, $\tan\theta$ increases rapidly to infinity.

## SOLVING TRIANGLES

If you know one of the acute angles of a right triangle, you know the other one—its *complement*.

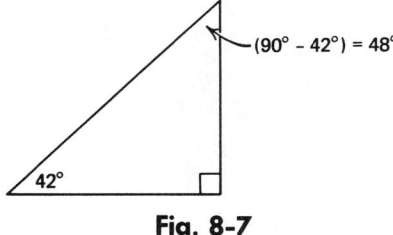

**Fig. 8-7**

If you know the length of one side of a right triangle and one of the acute angles, you can find the other two lengths.

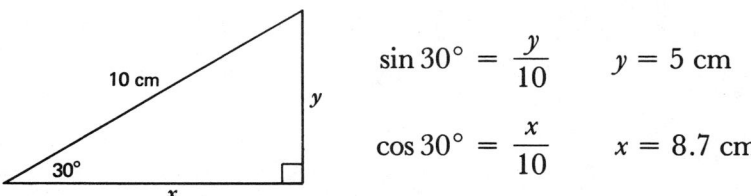

**Fig. 8-8**

Find *x* and the hypotenuse for this right triangle:

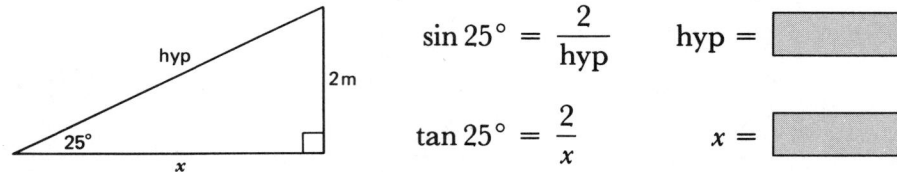

**Question 8-2.** If you know two sides of a right triangle, can you find the third side and the angles? Demonstrate:

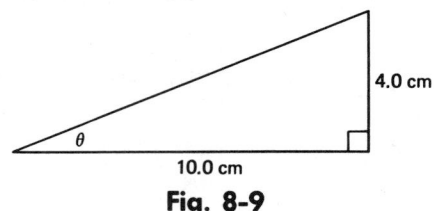
**Fig. 8-9**

If you know one of the acute angles of a right triangle (and therefore know the other acute angle) but do not know any of the side lengths, you cannot find the lengths without further information.

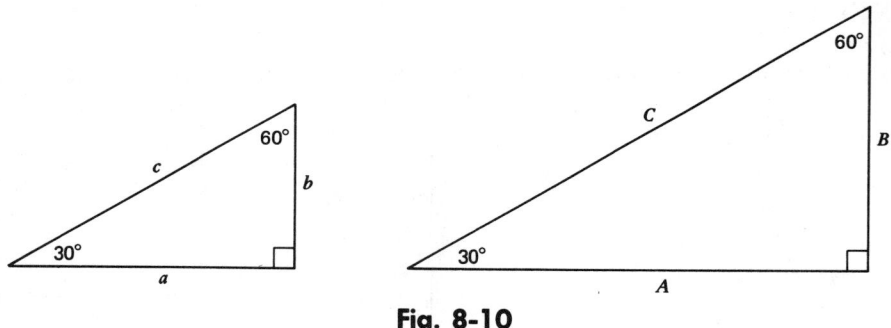
**Fig. 8-10**

The triangles in Fig. 8-10 are called *similar*. They have the same angles, but have different lengths. However, there are very useful relationships between the *ratios* of lengths of similar triangles:

$$\frac{b}{c} = \frac{B}{C} \qquad \frac{a}{b} = \frac{A}{B} \qquad \frac{a}{c} = \frac{A}{C} \qquad \frac{c}{C} = \frac{a}{A} = \frac{b}{B}$$

**Question 8-3.** Demonstrate this feature of similar right triangles by using the definitions of sine, cosine, and tangent.

Triangles can be similar without being right triangles. As long as each of the three angles in one triangle is equal to the corresponding angle in the other triangle, the equalities of length ratios are valid.

# SMALL ANGLE APPROXIMATIONS

In Fig. 8-11 we have drawn a construction that shows both $\sin\theta$ and a definition of $\theta$ itself. These relationships are

$$\sin\theta = \frac{\text{opp}}{R} \approx \frac{\text{arc}}{R} = \theta$$

For small $\sin\theta$, the opposite side of the triangle is almost equal to the arc length. Consequently, $\sin\theta \approx \theta$.

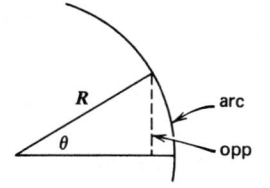

$$\sin\theta = \frac{\text{opp}}{\text{hyp}} = \frac{\text{opp}}{R}$$

but for small $\theta$, opp $\approx$ arc

**Fig. 8-11**

**Question 8-4.** Does this approximation claim that $\sin 30° \approx 30°$? How can this be when we know that $\sin 30° = \frac{1}{2}$?

Since $\tan\theta = \text{opp/adj}$, it is also true that $\tan\theta \approx \theta$ for small $\theta$. Let's see how good these approximations are. In the table below, fill out the actual values for $\sin\theta$ and $\tan\theta$ for the given angles as well as the values given by the approximation.

|     | $\theta$ | $\sin\theta$ | $\tan\theta$ |
|-----|----------|--------------|--------------|
| 0   |          |              |              |
| 1°  |          |              |              |
| 10° |          |              |              |
| 20° |          |              |              |
| 30° |          |              |              |
| 45° |          |              |              |
| 60° |          |              |              |
| 90° |          |              |              |

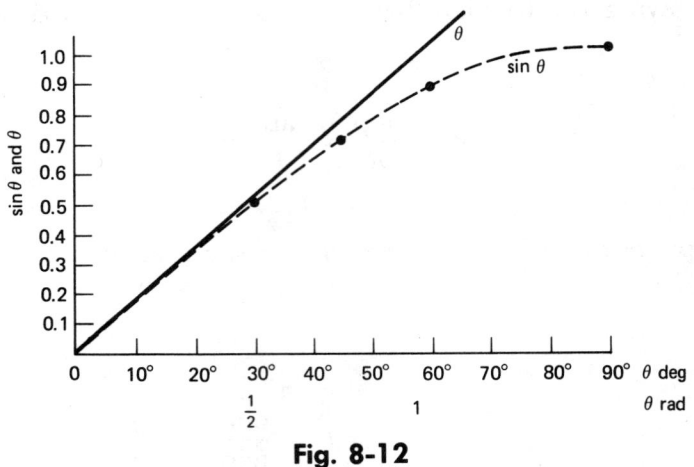

**Fig. 8-12**

In Fig. 8-12, we have plotted both $\sin\theta$ and $\theta$ as a function of $\theta$. Note that to plot $\theta$ versus $\theta$ we have expressed $\theta$ in radians. The graphs indicate why the small angle approximation works as well as it does for angles under 30°, and also shows how and why the approximation fails for larger angles.

## SURVEYING TECHNIQUES

One of the original uses of trigonometry was in land surveying. If you can measure one leg of a right triangle, and one of the acute angles (an angle less than 90°), then you can find the other two sides of the triangle and, of course, the third angle. Suppose, for instance, that a surveyor wants to measure the distance across a river where he cannot easily lay a measuring line. The arrangement is shown in Fig.

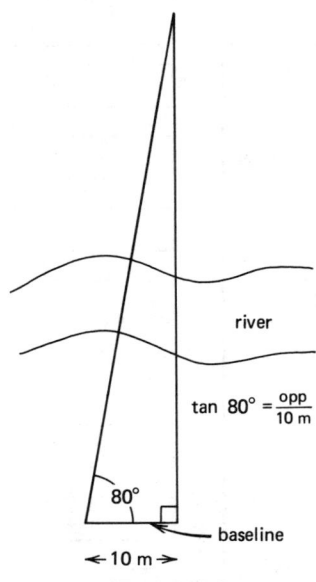

**Fig. 8-13**

8-13. The surveyor measures a baseline on his own side, and then sights at a marker across the river from each end of the baseline. In our example, we assume that in his first sighting he has set up a 90° angle between the baseline and the line to the marker. From the other end of the baseline, he sights to the marker and then measures the angle to the post at the original end of the baseline. Since $\tan \theta = \text{opp/adj}$ and since $\theta$ and the adjacent side are known, the opposite leg can easily be calculated. For instance, if the baseline is 10 m and $\theta = 80.0°$, then $\tan 80.0° = 5.67 = \text{opp}/10.0$ m. The distance across the river is therefore equal to 56.7 m.

## SMALL ANGLE SURVEYING

We can use surveying techniques to measure the distances to the nearest stars. For a long baseline, we can choose the diameter of the earth's orbit around the sun. If pictures are taken of the same section of the sky at six-month intervals, most of the stars will appear in the photographs at exactly the same positions with respect to each other. These are the so-called "fixed stars." A few stars, however, are close enough so that they will appear to shift their positions slightly in photographs taken six months apart. The geometry of this shift is shown in Fig. 8-14 in a greatly exaggerated form. The effect is called parallax. You can see the effect for yourself by holding your finger up at arm's length and viewing it first with one eye and then with the other eye. Compared with a distant wall, your finger will appear to shift position.

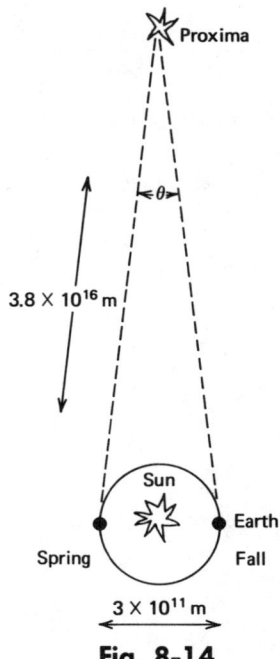

**Fig. 8-14**

In order to see what sort of angles we're dealing with in this kind of surveying, let's work the problem backward. The distance from our sun to the nearest star, Proxima, is $3.8 \times 10^{16}$ m. The diameter of the earth's orbit is $3 \times 10^{11}$ m. Using the definition of angles in radians, find the angle subtended by the earth's orbit at the star (the geometry is shown in the diagram). $\theta = $ _____ rad = _____ deg. With an angle this small, there is no need to worry about whether a right triangle should have been constructed and $\tan \theta$ calculated. Clearly, $\tan \theta \approx \theta$ for such a small $\theta$.

Incidentally, the displacements of stellar images on the photographic plates are so small that they must be measured with a microscope. Proxima is about 4 light-years away from us. The parallax method of measurement is good only out to about 100 light-years. The stars at greater distances have parallaxes too small to be measured precisely.

## TRIG IDENTITIES AND RELATIONSHIPS

Frequently in physics derivations, it is convenient to substitute one combination of trig functions for another. Here is a collection of the most commonly used substitutions and trig identities:

1. For any angle, $\sin^2\theta + \cos^2\theta = 1$. Prove this for yourself. Remember the definitions for sine and cosine in terms of the sides of a right triangle and use the Pythagorean theorem.

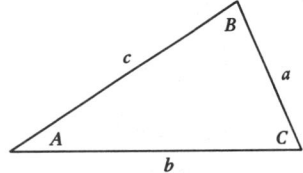

2. The sum of the interior angles of every triangle is 180°.

3. $\sin(\theta \pm \phi) = \sin\theta \cos\phi \pm \cos\theta \sin\phi$

   $\cos(\theta \pm \phi) = \cos\theta \cos\phi \mp \sin\theta \sin\phi$

   $\sin 2\theta = 2\sin\theta \cos\theta$

   $\cos 2\theta = \cos^2\theta - \sin^2\theta = 1 - 2\sin^2\theta = 2\cos^2\theta - 1$

   $\tan 2\theta = \dfrac{2\tan\theta}{1 - \tan^2\theta}$

4. $\sin\theta + \sin\phi = 2\sin\tfrac{1}{2}(\theta + \phi)\cos\tfrac{1}{2}(\theta - \phi)$

   $\sin\theta - \sin\phi = 2\cos\tfrac{1}{2}(\theta + \phi)\sin\tfrac{1}{2}(\theta - \phi)$

   $\cos\theta + \cos\phi = 2\cos\tfrac{1}{2}(\theta + \phi)\cos\tfrac{1}{2}(\theta - \phi)$

   $\cos\theta - \cos\phi = -2\sin\tfrac{1}{2}(\theta + \phi)\sin\tfrac{1}{2}(\theta - \phi)$

5. $\sin\theta \sin\phi = \frac{1}{2}\cos(\theta - \phi) - \frac{1}{2}\cos(\theta + \phi)$

   $\cos\theta \cos\phi = \frac{1}{2}\cos(\theta - \phi) + \frac{1}{2}\cos(\theta + \phi)$

   $\sin\theta \cos\phi = \frac{1}{2}\sin(\theta + \phi) + \frac{1}{2}\sin(\theta - \phi)$

6. Law of sines:

$$\frac{a}{\sin A} = \frac{b}{\sin B} = \frac{c}{\sin C}$$

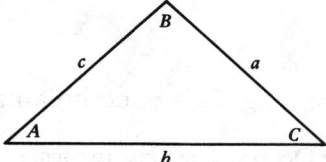

7. Law of cosines:

   $c^2 = a^2 + b^2 - 2ab\cos C$

   $b^2 = a^2 + c^2 - 2ac\cos B$

   $a^2 = b^2 + c^2 - 2bc\cos A$

8. cotangent $\theta = \cot\theta = \dfrac{1}{\tan\theta} = \dfrac{\cos\theta}{\sin\theta}$

   secant $\theta = \sec\theta = \dfrac{1}{\cos\theta}$

   cosecant $\theta = \csc\theta = \dfrac{1}{\sin\theta}$

## Answers to Questions

8-1. It appears from the diagram that there are a little more than 6 rad in the complete circle.

8-2. The hypotenuse could be found by the Pythagorean theorem:

   $c^2 = a^2 + b^2 \quad c^2 = 10^2 + 4^2 = 116 \quad$ hypotenuse $= c = 10.8$ cm

   Also: $\tan\theta = \dfrac{4}{10} = 0.4 \quad \theta = 21.8°$

   And: $\sin 21.8° = \dfrac{4.0}{hyp} \quad$ hyp $= 10.8$ cm

8-3. Note that $\sin 30 = b/c$ and also $B/C$.

   Therefore,

$$\frac{b}{c} = \frac{B}{C} \quad \text{and} \quad \frac{b}{B} = \frac{c}{C}, \text{ etc.}$$

8-4. The definition of angle $\theta = arc/r$ is in terms of *radians*. $\sin 30° = \sin(30/57 \text{ rad}) = \sin(0.53) = 0.5$.

   $0.53 \approx 0.5$

## PROBLEMS

1. In a right triangle, one of the acute angles is 15°. The hypotenuse is 15 cm long. What is the length of the side opposite the 15° angle? Adjacent to it?

2. One side of a right triangle is 25 m long. It is opposite an angle of 37°. What is the length of the hypotenuse? Of the other side?

3. One side of a right triangle is 150 m long. It is opposite an angle of 1.0°. What are the lengths of the other two sides of the triangle? (Do not use trig tables!)

4. What angle is subtended when you look at a clock that has a diameter of 30 cm and is on a wall 20 m away?

5. A right triangle has a hypotenuse of 1.5 cm, and one side has a length of 0.5 cm. What are the values for the angles and the third side?

# CHAPTER 9
# GEOMETRY AND DIAGRAMS

In studying physics, as with any other subject, style is very important. Some people work hard, spend inordinate amounts of time, and still can't solve problems, particularly on tests. A few people, however, seem to develop a knack for analyzing problems and finding the solutions. Almost always these few have learned to set up problems properly before attacking them. Some of this ability consists of very simple skills and habits that anyone can learn. There is no guarantee that if you follow these rules you will be able to solve every problem, but at least your work will appear to be organized. For practical purposes, that's worth extra credit.

For most physics problems, *draw a diagram first*. Sketch it carefully, roughly to scale, and label everything. If information given in the problem is not on your diagram, you probably haven't done it right.

Drawing a diagram need not take long. The very process presents you with a mechanical step-by-step chore that helps to organize the information and also prevents panic.

Some people think that they can't draw a straight line, and therefore can't draw diagrams. Nonsense! No artistry is required. With a bit of practice anyone can learn to sketch. The investment is minimal; the payoff is great.

Let's start with a square. Not like this:

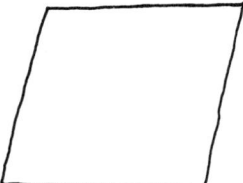

Keep the sides vertical, horizontal, and equal in length. No need to use a ruler.

Try some yourself →

Now draw some right triangles. First, one with a 45° angle.

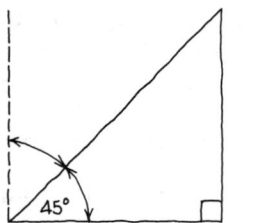

Note that the hypotenuse is half-way between vertical and horizontal.

Now draw some 30°–60°–90° triangles.

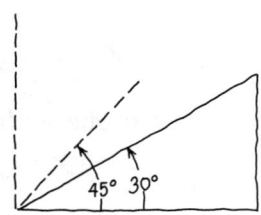

The 30° has to be $\frac{1}{3}$ of the 90°. Take the time to do it right. Without using a protractor, sketch a 10° right triangle.

The 10° is about $\frac{1}{4}$ of the 45°.

In physics, things are always happening to "objects." Objects are almost always cubes, or rectangular boxes, or cylinders, or spheres. Sometimes it helps to be able to sketch them in three dimensions. In general, keep your lines or planes vertical, or horizontal, or at an angle of 45° to the vertical.

Here is a box:                              Try a couple:

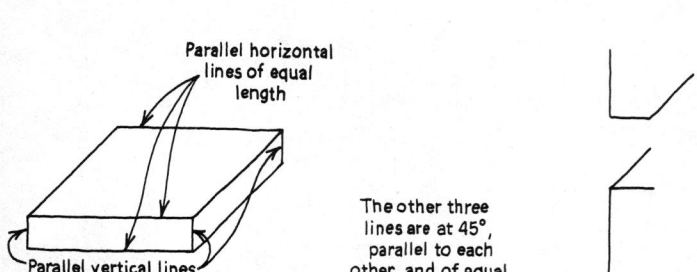

Cylinders usually have one oval end visible, straight sides parallel to each other, and half an oval visible at the other end.

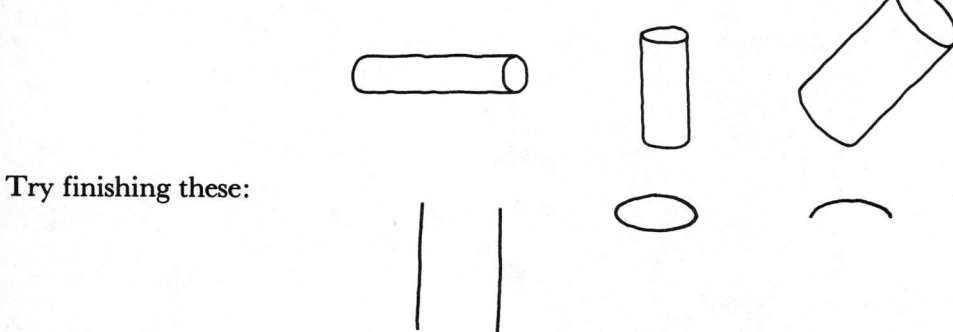

Try finishing these:

Spheres are hard to draw in three dimensions. They usually look like circles. You indicate the curved surface with a few curved lines:

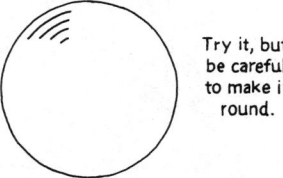

Alternatively, it is frequently useful to sketch an "octant" of a sphere.

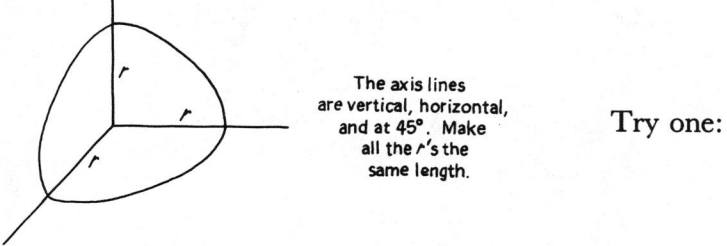

Try one:

In an octant, a radius to the spherical shell has projections on the horizontal plane and on the vertical axis:

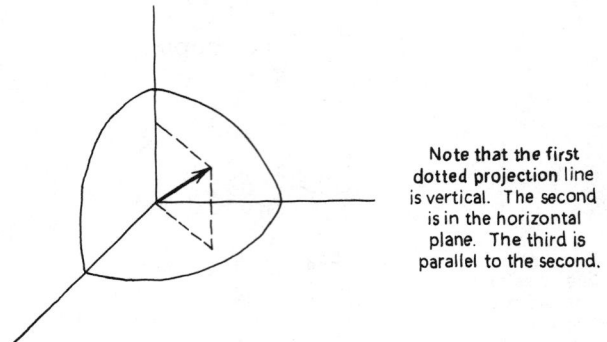

Note that the first dotted projection line is vertical. The second is in the horizontal plane. The third is parallel to the second.

## DRAWING AXES

In the chapter on graphs (Chapter 7), we claimed that data should always be graphed while they are being obtained. Anyone can learn to sketch a crude but satisfactory graph, even if no graph paper is available. No ruler is necessary. Do it freehand. Mark the units by eye if necessary.

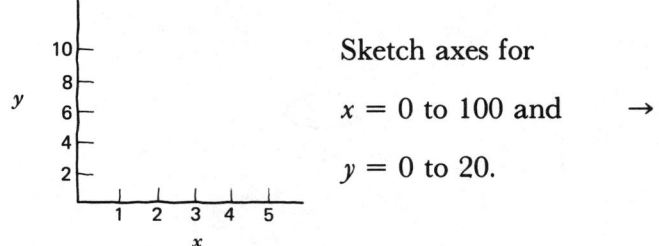

Sketch axes for

$x = 0$ to 100 and     →

$y = 0$ to 20.

The axes for three dimensional graphs should be labeled according to the right-hand-screw rule. If you turn the $x$ axis toward the $y$ axis, the $z$ axis is in the direction that the screw would advance if similarly turned.

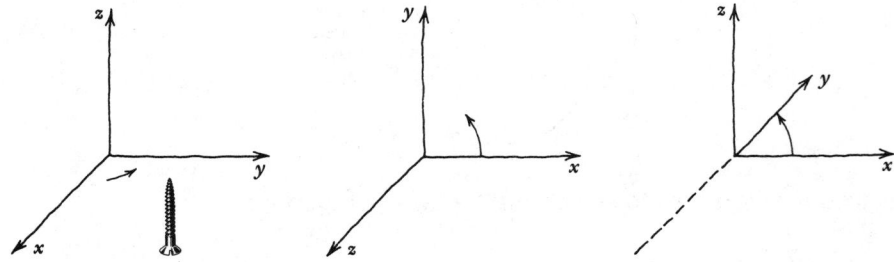

A position on a plane can be given in terms of two coordinates. They can be $x, y$, or $r, \theta$:

$x = r \cos \theta$

$y = r \sin \theta$

$r = \sqrt{x^2 + y^2}$

In three dimensions a point can be located at $x, y, z$, or at $r, \theta, \phi$.

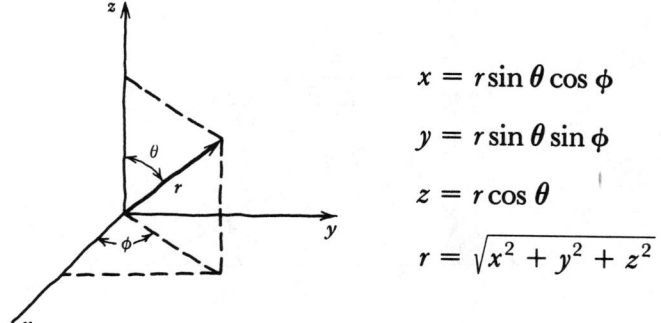

$$x = r\sin\theta\cos\phi$$
$$y = r\sin\theta\sin\phi$$
$$z = r\cos\theta$$
$$r = \sqrt{x^2 + y^2 + z^2}$$

## SOME NECESSARY GEOMETRICAL FACTS

### The Pythagorean Theorem

In a right triangle, $c^2 = a^2 + b^2$. Here is the proof.

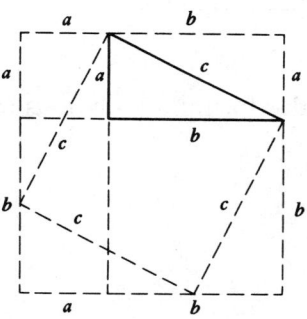

Complete the squares on the sides $a$, $b$, and $c$. Extend the sides to form a square as shown. Consider the areas of the squares, rectangles and triangles.

$$a^2 + b^2 + ab + ab = c^2 + \tfrac{1}{2}ab + \tfrac{1}{2}ab + \tfrac{1}{2}ab + \tfrac{1}{2}ab$$

area of large square = area of central square + 4 triangular regions

Therefore: $a^2 + b^2 = c^2$

### Law of Cosines

For any triangle (not necessarily a right triangle):

$$c^2 = a^2 + b^2 - 2ab\cos C$$

Note that if $\theta = 90°$, this formula reduces to the Pythagorean theorem.

# Law of Sines

For any triangle,

$$\frac{a}{\sin A} = \frac{b}{\sin B} = \frac{c}{\sin C}$$

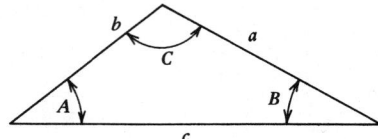

The distance between two points

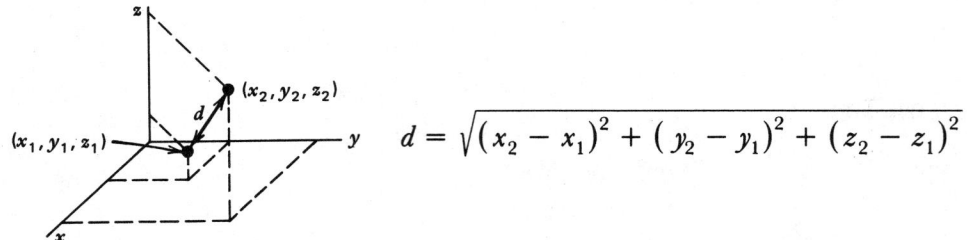

$$d = \sqrt{(x_2 - x_1)^2 + (y_2 - y_1)^2 + (z_2 - z_1)^2}$$

In two dimensions, this formula is obvious from the Pythagorean theorem.

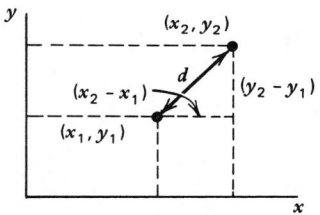

## Areas and Volumes of Common Shapes

| | | |
|---|---|---|
| Rectangle | area = $LW$ | perimeter = $2(L + W)$ |
| Triangle | area = $\frac{1}{2}hb$ | $h$ = height |
| | | $b$ = base |
| Circle | area = $\pi r^2$ | circumference = $2\pi r$ |
| | = $(\pi/4)d^2$ | = $\pi d$ |
| Sphere | surface area = $4\pi r^2$ | volume = $(4/3)\pi r^3$ |
| | = $\pi d^2$ | = $(\pi/6)d^3$ |
| Cylinder | curved surface area = $2\pi rh$ | volume = $\pi r^2 h$ |

Note that *all* areas contain length squared. Note that *all* volumes contain length cubed. Note that the **area of a circle** $[(\pi/4)d^2]$ is about $\frac{3}{4}$ of the area of a square in which it would just fit.

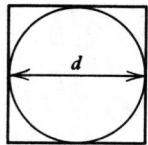

Note the volume of a sphere $[(\pi/6)d^3]$ is about $\frac{1}{2}$ of the volume of a cube in which it would just fit.

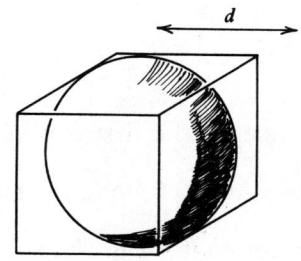

# CHAPTER 10
# MOTION AND THE POWER FUNCTIONS: $y = kx^n$

Almost any phenomenon that you will meet in the first two years of physics can be described with one or a combination of four simple mathematical functions. You can conclude that the world is very simple or else that introductory physics limits itself to the simple aspects of the world. In either case, most of your problems can be solved if you are familiar with the *power function*, the *sinusoidal function*, *exponentials*, and *logarithms*.

In each of these functional relationships, we will be concerned with only two variables at a time. The position of an object may be a function of time; the pressure of a gas may be a function of the temperature; the potential energy of an object may be a function of its position. By "function of" we mean that for a given value of one variable, such as the time, there is a particular value of the other variable, such as the position. Often one variable may depend on several others. For instance, the pressure of a gas depends on volume and temperature. For our purposes, however, we will consider only two variables at a time, one depending on the other. If there are other variables that might make a difference, we will keep them constant.

Although there are only four classes of these simple functions, each one has several different members or formulations. We will deal with these in turn, but always in terms of physical phenomena that require such description.

We have already seen that in taking measurements it is always necessary to make approximations. The same situation is often true in using simple functions to describe real-world phenomena. The orbits of the planets around the sun are actually ellipses. However, for a good first approximation, the orbits can be described as circles. The distance that a heavy object falls through air is approximately proportional to the square of the time, at least for the first few seconds. Then the approximation gets progressively worse as air friction changes the nature of the phenomenon. Radioactive atoms decay exponentially, but if the number of decays per time interval is small, then the data are subject to large random fluctuations.

First, we will study simple phenomena where the simple mathematical functions make good models. Later we will point out ways to use the simple functions as approximations to more complicated situations.

The family of power functions ($y = kx^n$) describes an enormous number of physical phenomena. No wonder! As you can see in Fig. 10-1, the graphs can look very different from each other, depending on the value of the exponent $n$. If $n = 1$, the function is linear; if $n = 2$, the function is quadratic, or the squared function. If $n$ is negative, we have the reciprocal functions. Let us deal with each in turn.

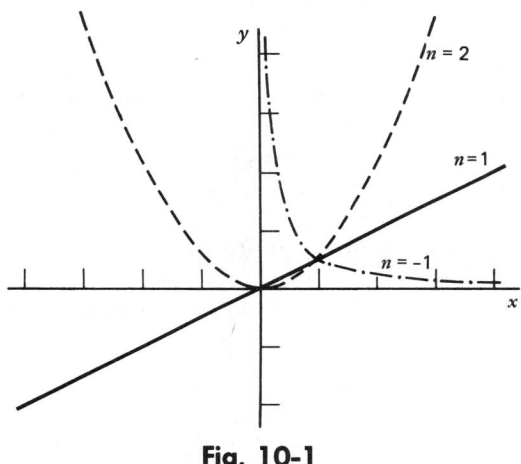

**Fig. 10-1**

## 1. THE LINEAR FUNCTION: $y = kx^1$

"*Double your pleasure, double your fun...*" (from an advertising slogan)

If we measure "fun" in terms of the time spent having it, then doubling your time might or might not double your pleasure. You can no doubt think of examples both ways. However, if doubling the time spent doubles your pleasure, you can claim correctly that the pleasure is *proportional* to the time:

$$\text{pleasure} \propto \text{time}$$

Note the symbol for "proportional to": $\propto$. If you double one variable, you also double the other. The definition of "proportional" is very specific; it does not mean simply that if one variable increases, so does the other. The increase must be such that if one variable is multiplied by some number, so is the other one.

Let's take an example. Suppose that you save money at the bank at the rate of $10 a week. At the end of the first week your account contains $10. At the end of the second week it has $20. Double the time; double the money. At the end of 52 weeks, you have $520. (We assume that no interest is being added.)

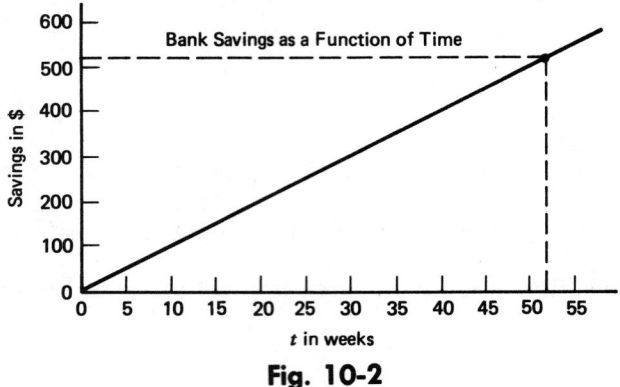

**Fig. 10-2**

The graph of your bank savings versus time is shown in Fig. 10-2. Note that the "$y$" axis is in terms of dollars and the "$x$" axis is in terms of number of weeks.

**Question 10-1.*** Does the proportionality definition hold for the entire time? Compare the amount you have at the end of 52 weeks with the amount you have at one-half that time (26 weeks).

A proportionality can be turned into an equation by using a "constant of proportionality":

$$\text{If } y \propto x$$
$$\text{then } y = kx$$

For instance, in the example of the savings program, one variable is the amount of money in the bank, $. The other variable is the time in weeks, $t$.

$$\$ \propto t$$

or

$$\$ = (\$10/\text{week})t$$

The constant of proportionality in this case is ($10/week).

---

*Our answers or comments on questions are at the end of the chapter.

## Nonproportionality, but Linear

Suppose that you start with $100 in your bank account and save at the rate of $10/week. Your savings are no longer *proportional* to the time. Look at the graph in Fig. 10-3. At the end of the first week you have $110; at the end of the second week you have $120. In double the time, your savings have only increased by $120/110 = 1.09$.

The equation describing this new situation is

$$\$ = (\$10/\text{week})t + \$100$$

In Fig. 10-3, the amount with which you started, the $100, is called the *intercept*.

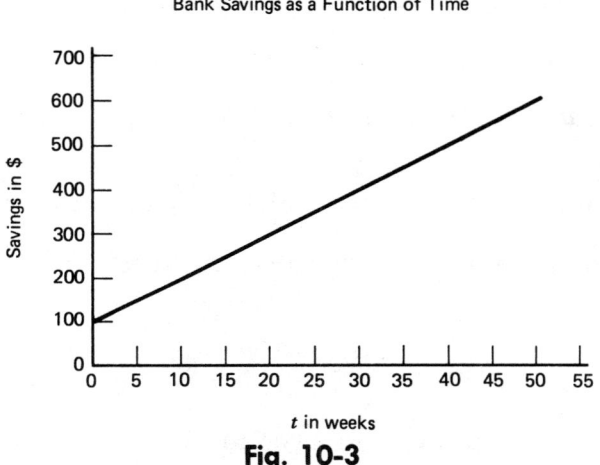

Bank Savings as a Function of Time

**Fig. 10-3**

## Linearity

The starting point of the relationship between two variables is often a matter of choice. For instance, Fig. 10-4 shows several relationships between distance ($x$) of a car from its starting point and time ($t$) since the clock started from zero.

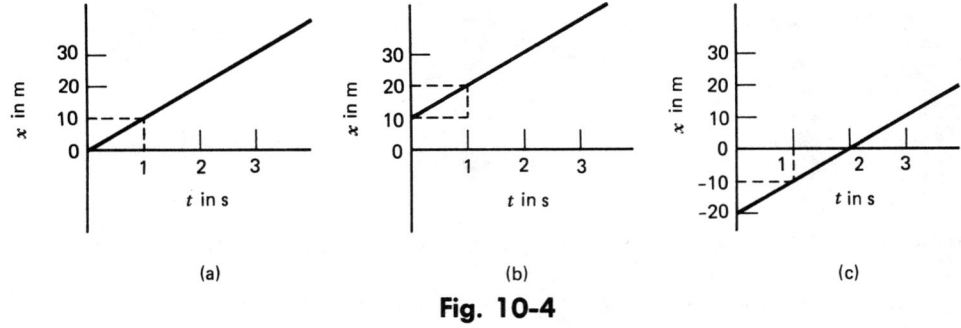

**Fig. 10-4**

Part (a) represents the motion of a car that started at $x = 0$ at $t = 0$, as shown in Fig. 10-5. Perhaps the car was directly in front of us, moving to the right, when we started the clock. Part (b) represents the motion of a car that was already down the road to the right when we started the clock. Graph (c) represents a car that did not pass our position until 2 seconds from the time we started the clock. At $t = t_0$, the car was to our left. If we wanted to, we could make all three motions start out at $x = 0$ at $t = 0$ by changing our own definition of $x$ so that the car is at the new origin at $t = 0$.

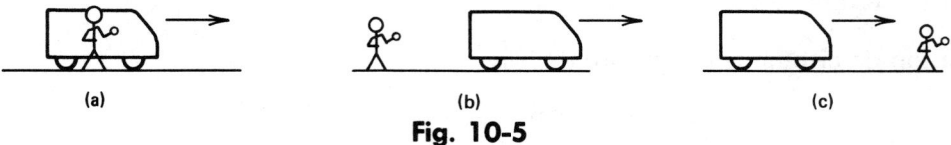

Fig. 10-5

Note that in all three cases, the car travels 10 meters in the first second. Its speed must be 10 meters per second (10 m/s). In fact, the car travels 10 m in *each* second. Its speed must be constant. In the graphs, this constancy is shown by the straight line relationship between $x$ and $t$. This is a *linear* relationship. Two variables can be linearly related without being proportional to each other.

**Question 10-2.** Is your age linearly related to the year date in this century? Is it possible for anyone's age to be proportional to the year date in this century?

## Slope of a Graph Line

In Fig. 10-6 we show the position graphs $x(t)$ for a car and a bike. They both passed our local $x = 0$ at our time $t = 0$. The car, evidently, is going faster than the bike. When $t = 1$ s, the $x$ position of the car is 10 m, but the bike has gone only 3 m. The car's speed is 10 m/s; the bike's speed is 3 m/s.

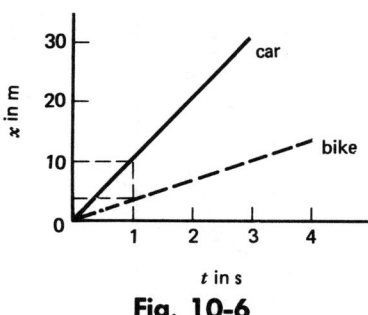

Fig. 10-6

In the graph of Fig. 10-6, the fact that each speed is constant is represented by straight lines. The *magnitude* of the speed is represented by the *slope* of each line. The steeper the slope, the greater the speed. We have a formal definition for the slope of a line on a graph:

$$\text{slope} = \frac{\text{rise}}{\text{step}} = \frac{\Delta x}{\Delta t}$$

[$\Delta$ = delta, a small interval of the variable. $\Delta x = (x_{\text{final}} - x_{\text{initial}})$]. For a straight line, the slope is constant. As shown in Fig. 10-8 the slope measurement can be made for any time interval $\Delta t$.

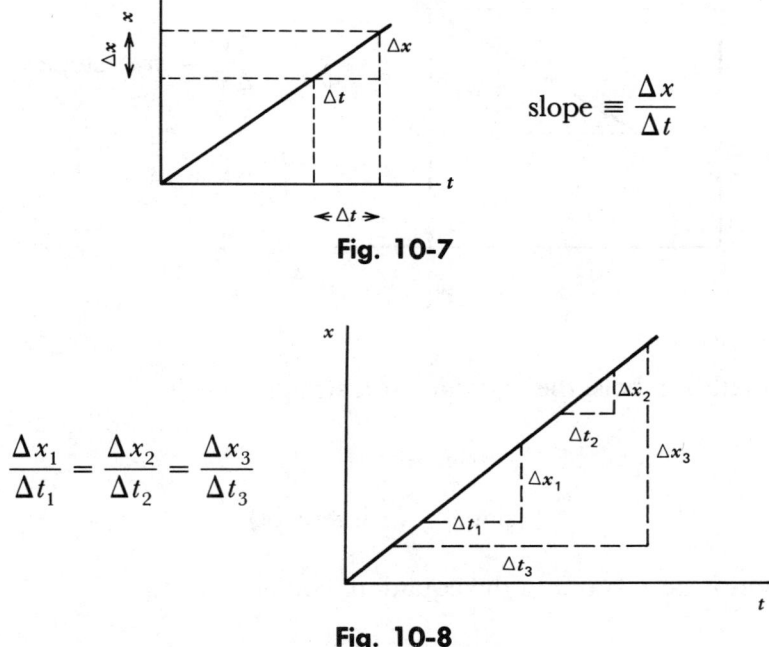

Fig. 10-7

$$\frac{\Delta x_1}{\Delta t_1} = \frac{\Delta x_2}{\Delta t_2} = \frac{\Delta x_3}{\Delta t_3}$$

Fig. 10-8

**Question 10-3.** Why can the slope measurement be made for any $\Delta t$? Is $\frac{\Delta x_1}{\Delta t_1} = \frac{\Delta x_3}{\Delta t_3}$? Actually measure the lengths and calculate the ratios.

## Equation of Linear Relationship

If $y$ and $x$ are linearly related, the slope of the graph of $y$ versus $x$ must be a straight line. Therefore, $\frac{\Delta y}{\Delta x}$ = constant.

In math texts, this constant representing the slope is traditionally called $m$. If $y$ is *proportional* to $x$, as shown in Fig. 10-9, then $y = mx$.

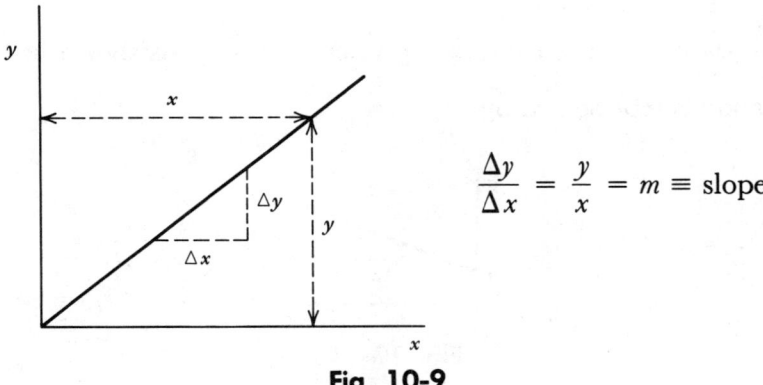

$$\frac{\Delta y}{\Delta x} = \frac{y}{x} = m \equiv \text{slope}$$

Fig. 10-9

However, if $y$ equals some initial value $b$ when $x = 0$, as shown in Fig. 10-10, then $y = mx + b$.

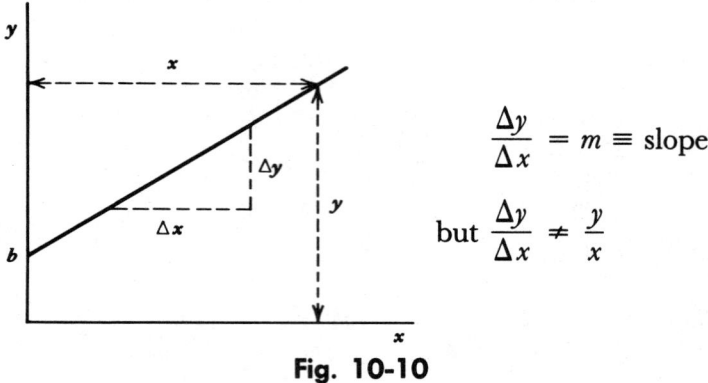

**Fig. 10-10**

$$\frac{\Delta y}{\Delta x} = m \equiv \text{slope}$$

but $\frac{\Delta y}{\Delta x} \neq \frac{y}{x}$

Using the traditional math symbols, the equation of a straight line is

$$y = \underset{\underset{\text{(slope)}}{\uparrow}}{m}x + \underset{\underset{(y \text{ intercept})}{\uparrow}}{b}$$

Note that when you substitute $x = 0$ into the equation, you get $y = b$.

**Question 10-4.** What does $x$ equal when $y = 0$?

## Constant Speed

Instead of $y$ and $x$, which represent any two variables, let's go back to $x$ and $t$. These describe motion along one direction—position as a function of time. The average speed, or velocity along the $x$ direction, is defined as $\bar{v} = \frac{\Delta x}{\Delta t}$. This definition matches our previous usage and everyday language: speed is measured in miles per hour, or meters per second.

**Question 10-5.** For future use, calculate equivalent speeds in different units:

60 mph = [ ] m/s    1 m/s = [ ] miles/h

For constant speed, the average velocity *is* the velocity: $\bar{v} = v$. As shown in Fig. 10-11 the linear relationship between $x$ and $t$ is represented by

$$x = x_0 + vt \qquad \qquad \frac{\Delta x}{\Delta t} = v$$

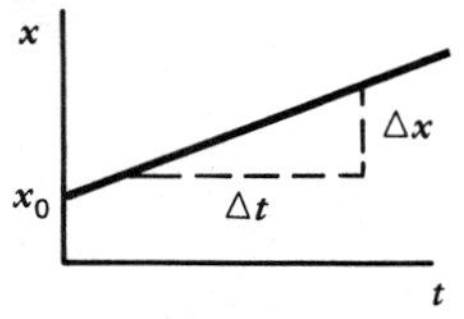

**Fig. 10-11**

**Question 10-6.** If you start out at 9:00 AM from Albany, which is 120 miles from New York City, and travel west at a nearly constant speed of 60 mph, how many miles (along the Thruway) are you from N.Y.C. at noon?

Let's ask another question about the previous problem. Given the equation and its graph, as shown in Fig. 10-12, where were you at 8:00 AM?

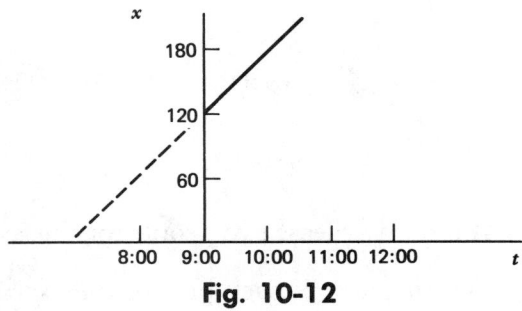

**Fig. 10-12**

One might be tempted to *extrapolate* the graph backward to "negative" time. In this case, one hour negative merely means 8:00 AM. Substituting $t = -1$ h into the formula yields

$$x = (60 \text{ mph})(-1 \text{ h}) + 120 \text{ miles} = 60 \text{ miles}$$

At 8 o'clock, according to the formula and the extrapolated graph, you were only 60 miles from New York City. However, as far as the original facts are concerned, at 8 o'clock that morning you may still have been eating breakfast in Albany! The math and the graphs are only models for reality. Like all models, they have limited domains of applicability. You may not be able to extrapolate beyond the original domain. We will see many other examples of the model nature of our math descriptions.

## Negative Slopes

In Fig. 10-13 we show a graph of the amount of money left as an inheritance. Once again, we assume that no interest is adding to the account. The equation for this linear relationship is

$$\$ = -(\$1000/\text{yr})t + \$10{,}000$$

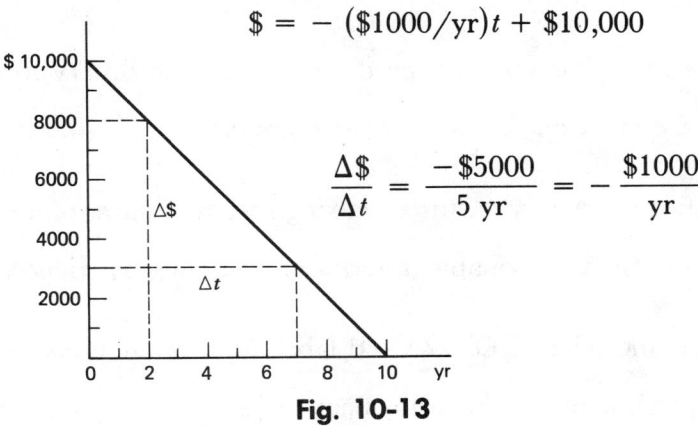

$$\frac{\Delta \$}{\Delta t} = \frac{-\$5000}{5 \text{ yr}} = -\frac{\$1000}{\text{yr}}$$

**Fig. 10-13**

**Question 10-7.** Check that the equation describes the graph by substituting values for $t = 0$, $t = 5$ yr, $t = 10$ yr, $t = -1$ yr, $t = 12$ yr.

The slope of this line *looks* negative and our definition agrees. As time *increases* ($\Delta t$ is positive), the amount of money left *decreases*.

$$\Delta \$ = (\$_{final} - \$_{initial}) = (\$3000 - \$8000) = -\$5000$$

**Handling the Phenomena** (A) You need at least two people for this exercise. A team of three or four is even better, provided that each participates in each part of the exercise. And there *is* exercise. You measure your own position as a function of time as you walk with several constant, but different speeds.

1. Mark off a course along a hall of at least 10 m—preferably as long as 50 m. Divide the course into at least five parts—preferably ten.

2. One person walks *at normal walking speed* along the course. Either the walker or observer calls off his passage past the marks. Someone else notes the time at each call.

3. Record and *sketch* $x(t)$. Judge the uncertainty in each measurement and represent these errors with appropriate bars on the graph.

4. Draw the best straight line through the error bars and measure the slope. Find the average speed in m/s. Write the equation for your $x(t)$. (In your report, discuss whether or not it is appropriate to draw a straight line.)

5. Repeat the procedures for a fast walking speed and a slow speed. Try to make one speed about twice the normal and the other one about half the normal.

6. Calculate, to one significant figure, your three walking speeds in terms of miles per hour. Make sure that these calculated speeds are reasonable in terms of everyday experience.

7. Sketch your graph here and label it. Choose and label appropriate units so that all three data sets can fit on the one graph. Use reasonable error bars on each datum point.

(B) With most springs, the harder you pull, the more they stretch. That sounds like a power function. Perhaps the stretch is proportional to the pull.

$$\text{stretch} \propto \text{pull}$$

To find out whether or not this is the way springs really work, we must become more quantitative. We can measure the stretch in meters or centimeters (though we must be careful to measure just the stretch, not the whole length of the spring). To measure "pull" we should define units of force. In the standard SI system, the unit of force is the *newton*. Its symbol is N. A kilogram mass and the earth are pulled toward each other with a force of 9.8 N. In other words, the weight of the kilogram mass is 9.8 N. In the English system, the unit of force is the pound. 1 kilogram weighs 2.2 lb. Therefore, 1 lb = 4.45 N. One newton equals 0.22 lb, or about 3.6 ounces (oz).

**Question 10-8.** What is your weight in newtons?

There are spring scales calibrated in newtons that we could use in our experiment to determine the behavior of springs. There would be circular reasoning involved, however; we would be using springs to investigate the nature of springs. Of course, we would be using calibrated springs with a manufacturer's assurance that the readings are correct. A simple way to escape this problem is to choose a bunch of identical objects which we can hang from a spring. We assume that if we hang two 1-kg masses from a spring, the force exerted by the gravitational field is just twice that exerted on 1 kg mass. (That assumption is not logically obvious, but let's assume it and keep going.) We can then measure the stretch of a spring as a function of the applied force by hanging successive numbers of equal weights from the spring.

How you do this experiment depends on what kind of spring you can get, and therefore on what king of hanging objects you need. If you get a very light spring, you may be able to use paper clips for the objects. If your spring is a rubber band you may want to use heavy nails. At any rate, choose enough objects so that the spring will be extended a sizable fraction of its length. Plot the force as a function of the stretch. Note two important points here. First, since you are changing the force and measuring the stretch, it would seem more reasonable to plot the stretch as a function of the force. The force would then be the independent variable. However, convention usually dictates that this particular graph appear with the force as a function of the stretch. We will see the reason for this in more detail later. Second, there is no need at this point to express force in units of newtons. If you are using paper

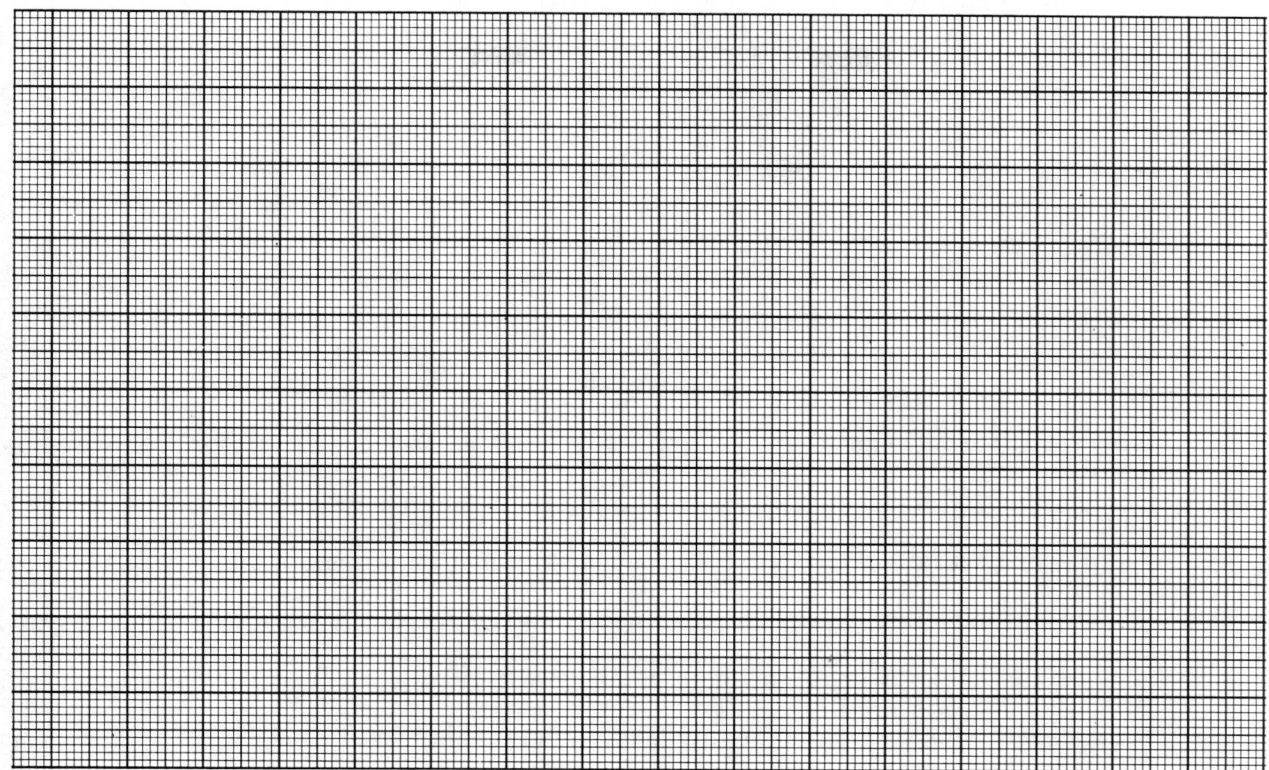

clips or nails, simply use their weight as arbitrary units. All we are after now is the functional relationship between force and stretch.

Should you expect to get a linear relationship or even a proportionality between force and stretch? If you use an ordinary coiled spring and don't stretch it too far, you may get a linear relationship. If you use a rubber band you will find that the graph curve soon departs from linearity. How about other springs? Consider the one that retracts the seatbelt in your car. If the force required to pull out that spring were proportional to the stretch, seatbelts would be lethal instruments. There are also springs that retract measuring tapes. Imagine how horrible it would be to have a 10-m measuring tape where the force required to pull it out would be proportional to the distance pulled! Evidently, only certain springs and under certain conditions have the property that $F = kx$. If springs follow this linear relationship, they are said to obey Hooke's law.

## 2. THE QUADRATIC, OR SQUARED, POWER FUNCTION: $y = kx^2$

The area of a square grows very rapidly as the length of each side increases. As a side increases from 1 to 2 to 3 to 4, the area increases from 1 to 4 to 9 to 16. The graph in Fig. 10-14 shows the area of a square as a function of the length of its side. The squared function is called "quadratic" because of this property of squares. ("Quad," as a prefix, means "four.")

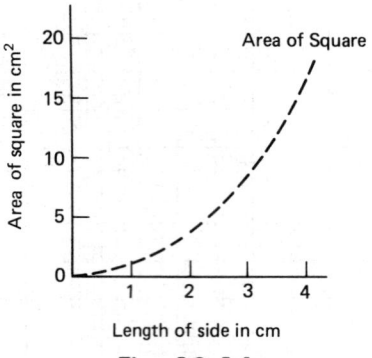

Fig. 10-14

Many other physical phenomena can be described by variables that depend on the square of some other variable. One such phenomenon is familiar, but you may never have noticed the details of the operation. If a heavy object falls freely from rest, the distance fallen is proportional to the square of the time of fall. You would think that it would be easy to demonstrate this fact by dropping something and timing its fall. The trouble is that dropped objects fall very rapidly and it's hard to measure what's going on. You can see this for yourself by timing the drop of an object from a height of 1 m, then from a height of 2 m. Use a stop watch if you can get one, or just try counting "one-thousand-and-one, one-thousand-and-two...." About how many seconds does the drop take from a height of 1 m? _____ About how many seconds does the drop take from a height of 2 m? _____ As you can see, with human timing methods precision is very poor. Still, check your data to see whether they are compatible with the square power function we suggested. For instance, did it take twice the time for twice the drop distance? Within the errors of your measurement, was the time taken for the 2-m drop equal to the square root of 2 times the time for the 1-m drop?

You can slow down the effect of gravity by rolling a marble down a very slight incline. To try such an experiment, you should have a very flat board, such as a good dining room table. Prop up one end slightly, so that a marble (or a toy cart) will roll faster and faster down the incline. Place markers where the ball is, at the end of 1 s, 2 s, and 3 s. In each case, measure the distance from the starting point, and see if your data agree with the formula $x = kt^2$.

Instead of slowing down the effect of gravity, we can increase the precision of our timing instrument. The photograph in Fig. 10-15 was taken with a strobe light that was flashing at exactly 30 times per second. The time exposure film shows the position of a ball bearing each $1/30$ s after it was dropped. You can read off the distance fallen on the meter stick in the picture. Take the data from the picture and plot it on the graph in Fig. 10-15. From your data, what is the best value for $k$ if the falling ball obeyed the equation $y = kt^2$? _____. (In plotting your data, remember to use error

# THE QUADRATIC, OR SQUARED, POWER FUNCTION: $y = kx^2$

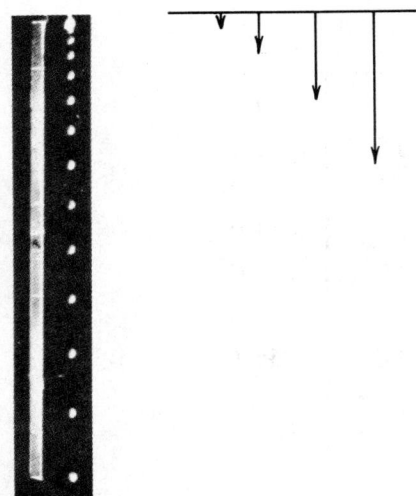
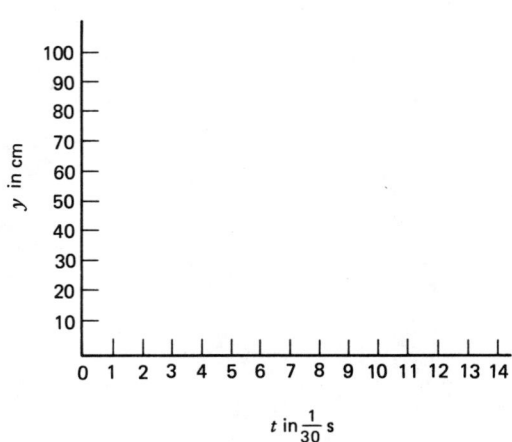

**Fig. 10-15**

bars to represent your uncertainties in the photograph measurement. You can assume that the errors in the strobe light frequency were small compared to the distance measurement.)

For constant speed, the slope of the $x(t)$ straight line was equal to the speed: $\frac{\Delta x}{\Delta t} = v$. Once again, with the quadratic function, we have a graph of $x(t)$, only this time it is not a straight line. How can we measure the slope of a curving line? Evidently, there is a different slope at every point. Geometrically, the slope at any point of a curve is equal to the slope of the tangent at that point. The mathematical operation is easy to perform on an actual graph with a ruler and pencil. As you can see in the diagram, place your pencil on the point where you want to know the slope, bring the ruler up to the pencil and swing it back and forth until it appears by eye to be tangent to the curve. With a little bit of practice you can get good precision from this method. Once you have drawn the tangent line, you can find its slope by the same method used in the previous section. The process is shown in Fig. 10-16.

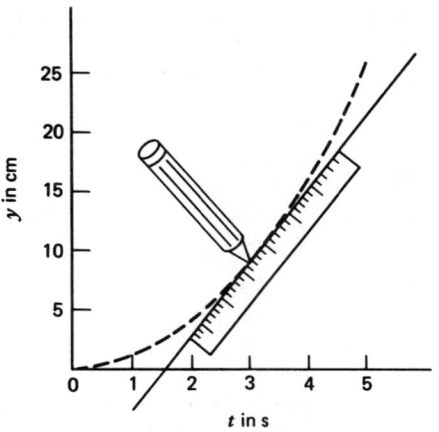

**Fig. 10-16**

# MOTION AND THE POWER FUNCTIONS: $y = kx^n$

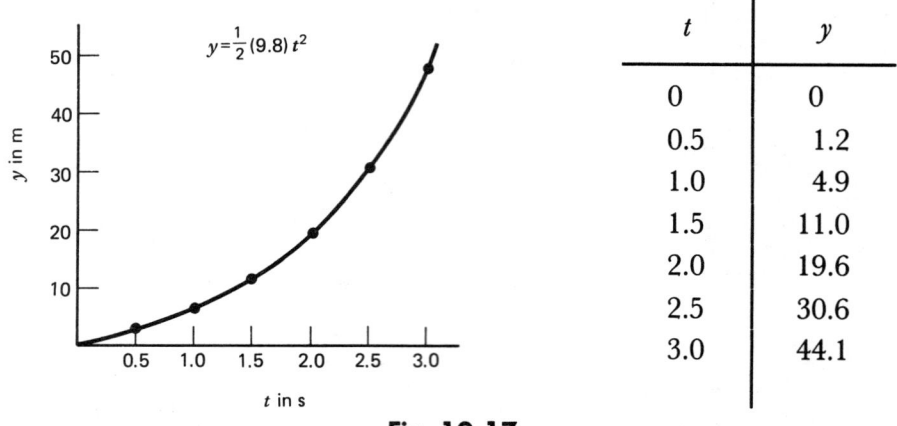

| $t$ | $y$ |
|---|---|
| 0 | 0 |
| 0.5 | 1.2 |
| 1.0 | 4.9 |
| 1.5 | 11.0 |
| 2.0 | 19.6 |
| 2.5 | 30.6 |
| 3.0 | 44.1 |

**Fig. 10-17**

Use this method to find the speed of a falling object at various times as shown on the graph in Fig. 10-17. This graph was not constructed from experimental data, but follows the actual formula for an object falling without air resistance. (This way you can have a smooth curve on which to practice taking tangents.)

Draw the tangents, and find the speeds at $t = 0, \frac{1}{2}, 1, 2, 3$ s.

| $t$ | Slope of Tangent in m/s |
|---|---|
| 0 | |
| $\frac{1}{2}$ | |
| 1 | |
| 2 | |
| 3 | |

Now that you have a new table of values for the speed as a function of time for a freely falling object, plot those values in the next graph. Join the data points with the best-fitting curve. According to your graph, what is the formula for the speed as a function of time for a freely falling object? _____.

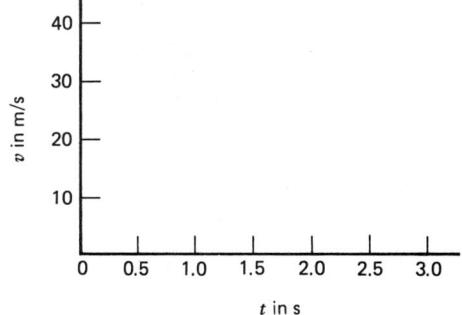

## Algebraic Derivation of Slope

It is often inconvenient and sometimes a little sloppy to draw tangents to a curve on a graph. There is an algebraic equivalent to the process. Remember that the slope of a straight line is defined to be equal to $\frac{\Delta y}{\Delta x}$. If the interval $\Delta x$ is centered around a particular point $x$, then $\frac{\Delta y}{\Delta x}$ appears as a chord on the graph as shown in Fig. 10-18.

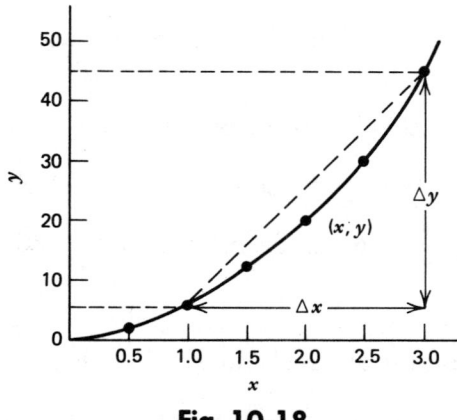

**Fig. 10-18**

If we shrink $\Delta x$ smaller and smaller about the point $x$, then the chord turns into the tangent. Let's do that algebraically for the case where $y = kx^2$.

$$\frac{\Delta y}{\Delta x} = \frac{\Delta(kx^2)}{\Delta x} = \frac{k(x + \Delta x)^2 - kx^2}{\Delta x}$$

$$= \frac{k(x^2 + 2x\Delta x + \Delta x^2) - kx^2}{\Delta x} = \frac{k2x\Delta x - k\Delta x^2}{\Delta x}$$

$$= 2kx - k\Delta x$$

As we shrink the interval $\Delta x$ to 0, the last term disappears. We have shown algebraically that the slope of the function $y = kx^2$ at the point $x$ is equal to $2kx$.

**Question 10-9.** Does this expression agree with the slopes that we found geometrically by drawing tangents?

Let's apply this formula for the slope to the special case of a freely falling object. If the distance fallen is $y$, then $y = kt^2$. For reasons that we will soon see, this formula is usually written $y = \frac{1}{2}gt^2$. (A new constant $\frac{1}{2}g$ has been substituted for $k$.) The slope of this curve at any point is equal to the speed at that point. Therefore, $v = \frac{\Delta y}{\Delta t} = 2kt = gt$. The speed of a freely falling object increases linearly with time.

Now let's apply the rules that we have developed to see how fast the speed is changing. The slope of the straight line $v = gt$ is $\frac{\Delta v}{\Delta t} = g$. The time rate-of-change of velocity $\frac{\Delta v}{\Delta t}$ is called the acceleration. In this case, the acceleration is equal to $g$ (instead of $2k$). Now we see why we substituted constants ($\frac{1}{2}g = k$). A freely falling object has a constant acceleration; its speed increases steadily. The value of $g$ at sea level is 9.8 m/s².

**Question 10-10.** Are the units and dimensions correct for acceleration?

The speed of a freely falling object starting from rest is given by $v = gt = (9.8 \text{ m/s}^2)t$. With that acceleration it doesn't take long for a falling object to gain high velocity. How fast would a golf ball be traveling after 2 s of fall? ▭ in m/s. Transform that speed into miles per hour. ▭ mph.

Actually, objects don't continue to fall freely in air. For dense human-sized objects the resistive force of air friction is proportional to the square of the speed. As the speed of fall increases, the resistive force increases quadratically (and therefore rapidly), until finally it is as large as the weight of the object. From that point on, the net force on the falling object is 0 and the object continues to fall at constant speed. A graph of the effect is sketched in Fig. 10-19. The falling object reaches a terminal speed. For a man with a parachute, that speed is about 15 mph. For a human without a parachute, the speed is about 120 mph. For a golf ball or a rock, the speed might get as large as 150 mph. As you can see, it does not take many seconds of falling for an object to approach its terminal speed and for the formula describing its fall to deviate from the simple formulas that we have derived.

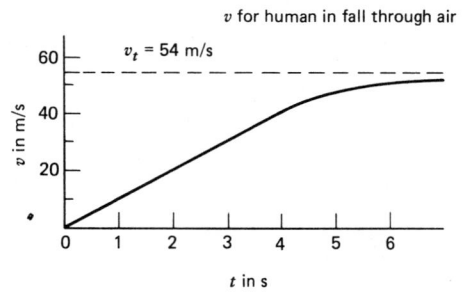

Fig. 10-19

In Fig. 10-20, we have summarized with graphs and formulas the functions relating position, speed, and acceleration for an object falling freely from rest. Note the formula for the distance fallen as a function of time. You started out this section by dropping an object from a height of 1 m, and then from a height of 2 m. Solve the formula to see how many seconds the fall should take in each case. Time to fall from 1 m: ☐ s. Time to fall from 2 m: ☐ s.

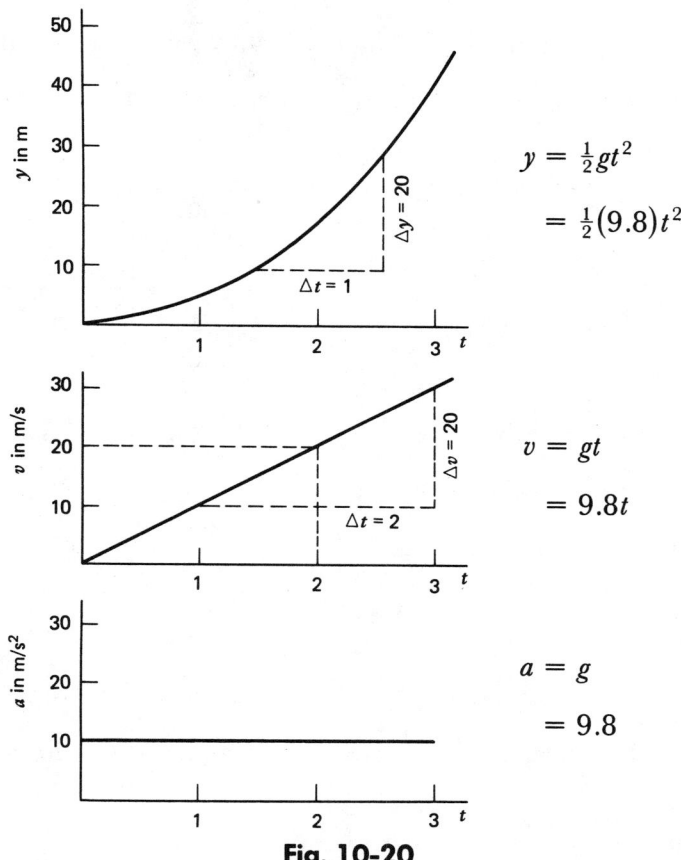

**Fig. 10-20**

The quadratic formula can actually be more complicated than the simple forms that we have used: $y = kx^2$ or $y = \frac{1}{2}gt^2$. The graphs of these simple forms pass through the origin, where $y = 0$ when $x$ or $t = 0$. We also want to be able to describe the motion of an object that is thrown down with an original speed. For instance, $y = v_0 t + \frac{1}{2}gt^2$.

**Question 10-11.** What is the speed of the object whose motion is described by this formula? Take $\frac{\Delta y}{\Delta t}$ as $\Delta t$ goes to 0.

The algebraic expression might get even more complicated. We have been using $y$ to indicate the distance fallen, presumably from some initial zero position. Now let's use position variables $x$ and $y$ in the standard method. To describe any motion, we choose axes for $x$, $y$, and $z$. Usually $x$ is in the horizontal plane, with $+x$ indicating the position to the right of the origin, and $-x$ indicating a position to the left. Either $y$ or $z$ is used for vertical position with $+$ meaning values above the origin and − meaning values below the origin. Let's rewrite our equation for the motion of a freely falling object and pay attention to these sign conventions. We'll place the origin, where $y = 0$, on the ground. Then if an object is dropped from some height $y_0$, the description is $y = y_0 + \frac{1}{2}gt^2$. *However*, the acceleration $g$ must now be given its proper direction. It is always down and therefore always negative: $g = -9.8 \text{ m/s}^2$. The formula becomes $y = y_0 + \frac{1}{2}(-9.8 \text{ m/s}^2)t^2$. The graph is shown in Fig. 10-21.

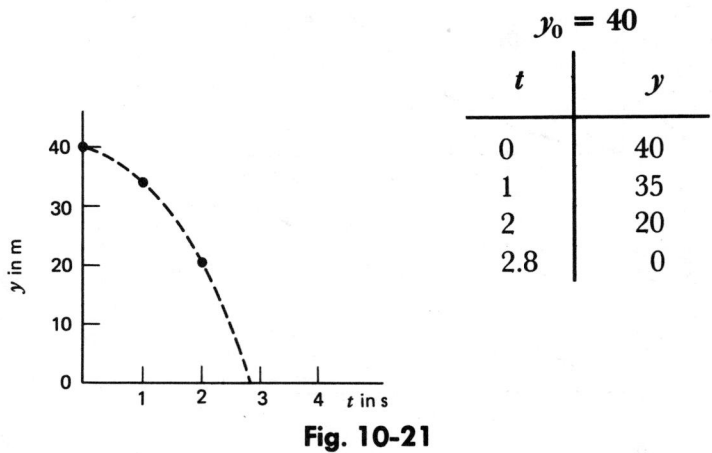

**Fig. 10-21**

If we give the object an initial velocity, we must assign it the proper direction. Suppose, for instance, that we throw the ball straight down from a height of 100 m with an initial speed of 5 m/s. The position of the object is given by

$$y = 100 \text{ m} + (-5 \text{ m/s})t + \frac{1}{2}(-9.8 \text{ m/s}^2)t^2$$

How long will it take the object to hit the ground? At $t = 0$, $y = 100$ m. What is the value of $t$ when $y = 0$? (For ease in solving the problem, use the approximation $9.8 \approx 10$.) ▢ s.

Note the simple and very important relationships between position, speed, and acceleration as functions of time. The slope of the position graph $x(t)$, at a given time, is equal to the speed at that time. The slope of the graph of speed as a function of time $v(t)$, at a particular time, is the acceleration at that particular time. Given a graph of $x(t)$ we could plot graphs of $v(t)$ and $a(t)$. That is just what we did in the graphs on page 83, showing the motion of an object dropping freely from rest. There is another relationship between these graphs. If you know the speed as a function of time when you start, then you can tell how far you have gone. For instance, in the simplest case, suppose that you are traveling at 60 mph for 2 h, as represented in the graph of Fig. 10-22. How far have you traveled? ▢ miles. Algebraically, you find the answer by using the formula $x = vt$. In this case, $v$ is constant and so the equation is valid. Graphically, you could find the answer by measuring the area

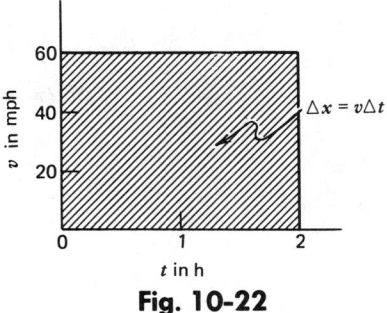

**Fig. 10-22**

under the line showing the speed from 0 to 2 h. That area is shown shaded by diagonal lines in Fig. 10-22.

**Question 10-12.** How can an area represent a distance traveled? Aren't the dimensions wrong?

It is generally true that the area under a curve of $v(t)$ is equal to the distance traveled. For instance, in Fig. 10-23 the speed is not constant, but is changing linearly with time (this is the case for freely falling objects). The area under the curve from 0 to 1 s is just equal to the area of a triangle which is $\frac{1}{2}$ base $\times$ height. In this case, the base is 1 s and the height when $t = 1$ s is 9.8 m/s. The distance fallen is, therefore, $\frac{1}{2}(1 \text{ s})(9.8 \text{ m/s}) = 4.9$ m.

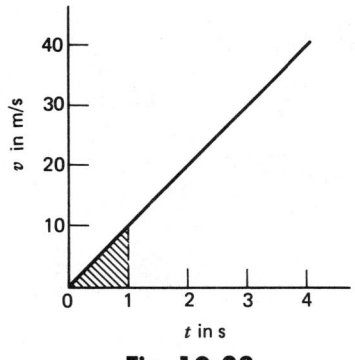

**Fig. 10-23**

For any ordinary motion, whether the $v(t)$ line is straight or curved, the distance traveled in an interval of time is equal to the area under the $v(t)$ curve bounded by that interval of time. In Fig. 10-24, an example is shown for a case of a ball that is thrown vertically downward with an initial velocity of 5 m/s. We can find the distance fallen during the first second by dividing the area up into a rectangle and a triangle. As shown in the diagram, the area of the rectangle is 5 m and the area of the triangle is $\frac{1}{2}(1 \text{ s})(9.8 \text{ m/s}) \approx 5$ m. The total area, therefore, is approximately equal to 10 m.

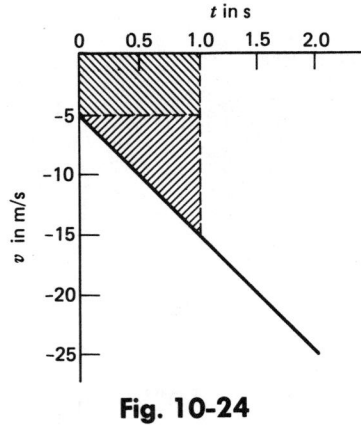

**Fig. 10-24**

What happens if the $v(t)$ line is not made up of straight lines? The area could then be found by counting small squares on the graph paper. The mathematical process of determining the area under curves is called "integration." In calculus classes, mathematicians teach many fancy methods for integrating both simple and complicated functions.

If we know the acceleration as a function of time, then in a similar way we can find the velocity as a function of time. The area under a curve of $a(t)$ is equal to the change of velocity during that time interval. For instance, at constant acceleration, we have $v = at$.

## Summary of Motion Graphs

The graphs of $x(t)$, $v(t)$, and $a(t)$ are all related, as shown in Fig. 10-25. The velocity at any given time has the value of the slope of the $x(t)$ curve at that time. This is because the average velocity during an interval $\Delta t$ is defined as $\bar{v} = \dfrac{\Delta x}{\Delta t}$. The instantaneous velocity at a particular time has the value of the tangent to the curve at that particular time: $v = \dfrac{\Delta x}{\Delta t}$ as $\Delta t$ goes to 0.

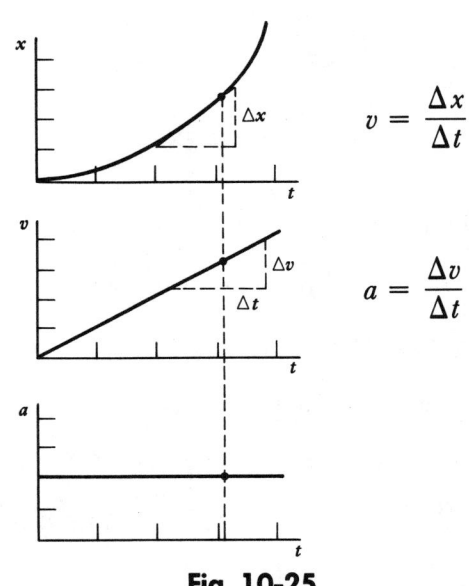

**Fig. 10-25**

The acceleration of an object at any given time has a value equal to the tangent of its $v(t)$ curve at that particular time. This is because $\bar{a} = \dfrac{\Delta v}{\Delta t}$. The instantaneous acceleration at a given time is $a = \dfrac{\Delta v}{\Delta t}$ as $\Delta t$ goes to 0. On a graph, the instantaneous acceleration has the value of the tangent to the velocity curve at that time.

Furthermore, the $x(t)$, $v(t)$, and $a(t)$ curves are related in terms of the areas under the curves, as shown in Fig. 10-26. For instance, the change of velocity during an interval of time is equal to the area under the $a(t)$ curve bounded by the interval of time. Similarly, the distance traveled during an interval of time is equal to the area under the $v(t)$ curve bounded by that interval of time.

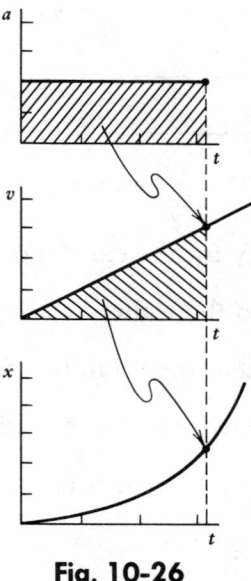

**Fig. 10-26**

Let's see how these general rules apply to an idealized journey of a car. Suppose the car is directly in front of you at $t = 0$. Its velocity during the following 4 s is shown in the $v(t)$ graph of Fig. 10-27.

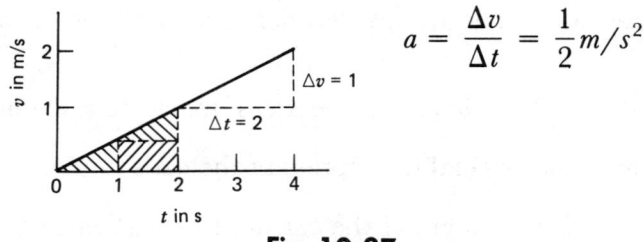

**Fig. 10-27**

The motion of the car can only be along the $x$ axis, although it can be either to the right (positive) or the left (negative). In this case, the car started with zero velocity at $t = 0$, and for the next 4 s kept going faster and faster to the right. Its acceleration is equal to the slope of the $v(t)$ curve. Since the slope is constant, the acceleration is constant. It has the value of $+\tfrac{1}{2}$ m/s². The distance traveled during the first second is equal to the area under the $v(t)$ curve during the first second. This triangular area is equal to $\frac{1}{4}$ m. During the next second, the area covered is $\frac{3}{4}$ m. The total distance traveled in 2 s is equal to 1 m. At the end of 4 s, the triangular area of the graph represents a distance traveled of 4 m.

Note that the constant acceleration gives rise to a velocity that is increasing linearly with the time, and a distance traveled that is increasing quadratically with time.

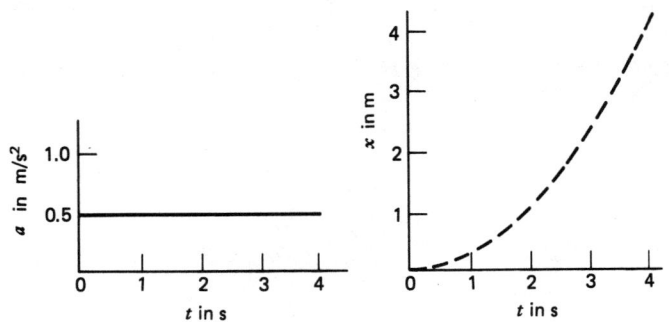

Now let's consider a different journey for the car. Once again, at $t = 0$ the car is directly in front of us. Its velocity, however, is positive (to the right). It has come from our left at high speed and we start timing it just as it passes us. From that point on, however, it is slowing down until at $t = 2$ s the car's velocity is 0. Meanwhile, it must have traveled considerably to our right. The car's speed then becomes more and more negative, implying that it is coming back toward us (to the left). Evidently, the driver jammed on his brakes, coming to a stop at $t = 2$ s, then threw the car into reverse and accelerated the car back toward us. At $t = 4$ s, the car is up to its original speed of 20 m/s, but is traveling to the left. Once again, the acceleration of the car is easy to calculate, since the slope of the $v(t)$ curve is constant. That slope is equal to $-10$ m/s$^2$.

**Question 10-13.** How can a constant acceleration slow a car down *and* speed it up again?

We can plot $x(t)$ for the car by calculating the areas under the $v(t)$ curve. Our results had better agree with common sense and our original description of the car's motion. First, we said that at $t = 0$, $x = 0$. From that time on, the displacement of the car must be further and further to the right for the first 2 s. At that time, the car turns around or backs up and comes back toward us. Its distance away from us gets smaller and smaller. Indeed, because of the symmetry, we ought to expect that the car will end up directly at the beginning point at 4 s. Taking areas under the curve, as shown in Fig. 10-28, during the first second, the car travels 15 m. During the next second, it travels an additional 5 m.

# THE QUADRATIC, OR SQUARED, POWER FUNCTION: $y = kx^2$

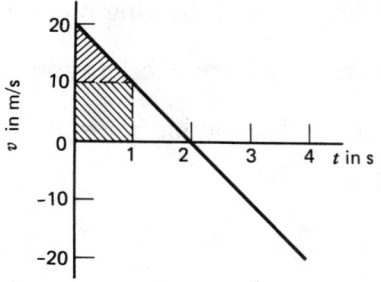
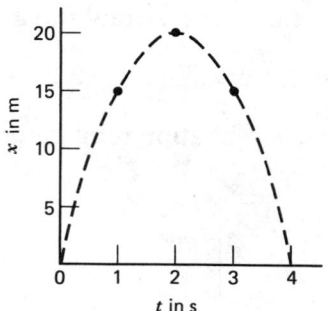

**Fig. 10-28**

Therefore, the turnaround point is 20 m to our right. During the third second, the area bounded by the curve and the axis is 5 m, but it is negative. A negative distance corresponds to a motion to the left. The car is coming back toward us. During the fourth second, the car travels an additional $-15$ m, bringing it directly past its starting point and heading to the left with a velocity of $-20$ m/s.

Our make-believe car did not make a trip that is physically very realistic. However, the same graph could describe very faithfully the actual vertical throw of a ball upward with an initial velocity of 20 m/s. Indeed of $x(t)$ representing motion along the horizontal $x$ axis, we could use $y(t)$, representing vertical motion. After 2 s, the ball has reached its turnaround point at 20 m, and is ready to start falling down. At 4 s, it is at its starting point and heading downward with a velocity of $-20$ m/s. (For ease of arithmetic, we have made the approximation that $g = -9.8$ m/s$^2 \approx -10$ m/s$^2$.)

Follow the same procedures with the next set of graphs to determine the $x(t)$ and $a(t)$ curves, assuming that at $t = 0$, we have $x = 0$.

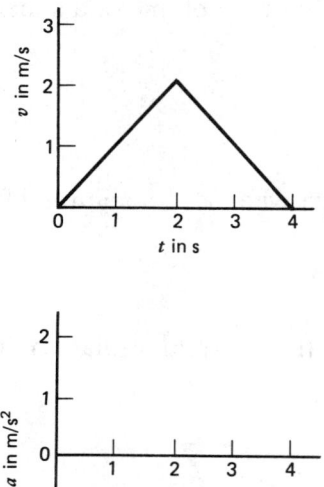
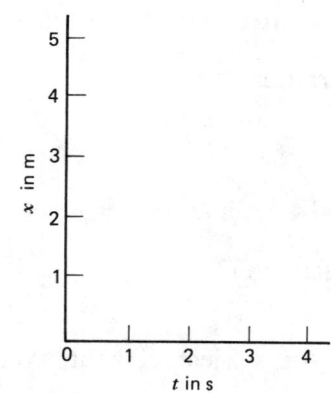

In the next set of curves, draw the graphs corresponding to the following motion: At $t = 0$, a car passes you at $x = 0$, traveling at $-2$ m/s. It slows down with constant acceleration so that at $t = 2$ s, you get $v = 0$. The acceleration remains constant so that at $t = 4$ s, you have $v = +2$ m/s.

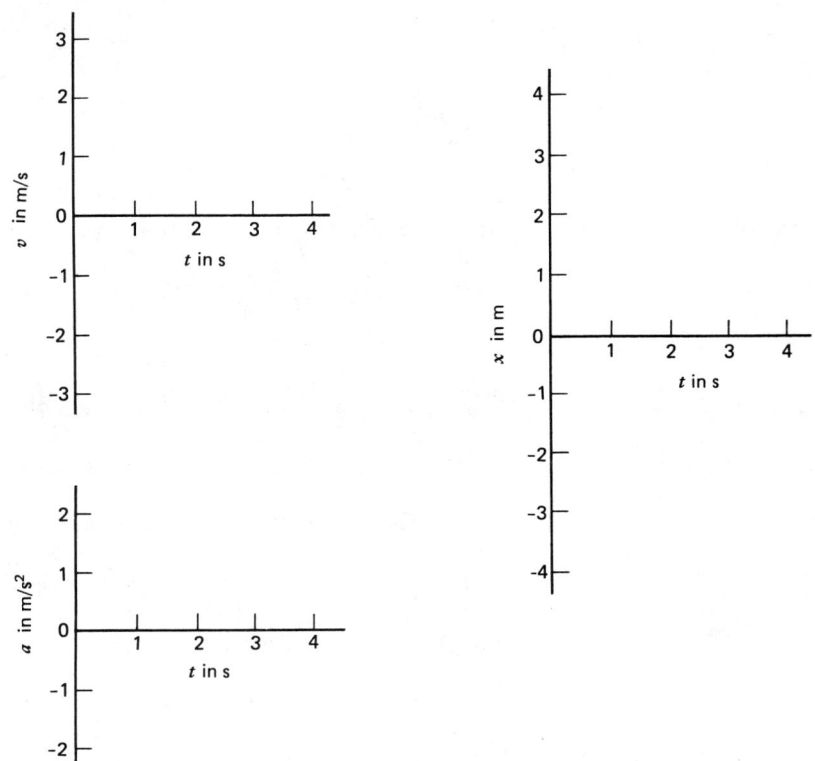

## 3. THE CUBIC: $y = kx^3$

In introductory physics, you meet only a few phenomena that must be described by power functions with an exponent greater than 2. One of these exceptions is that the volume of a sphere is proportional to the cube of its radius:

$$V_{\text{sphere}} = \tfrac{4}{3}\pi r^3$$

**Question 10-14.** Since the radius is half the diameter, what is the volume of a sphere in terms of its diameter?

When you study the subject of heat, you will learn that the thermal radiation from an object is proportional to the fourth power of its temperature: $E \propto T^4$.

**Handling the Phenomena** The rapid increase of the volume of a sphere as the radius increases can be quite startling if you do not expect it. Collect three or four spheres of different sizes and measure their diameters and volumes. The diameters can be measured accurately enough by sighting the sphere against a ruler. Better yet, lay a ruler flat on a table, put the ball on it, and butt rectangular shapes, like books, up against the ball. The easiest way to measure volume is by water displacement. Either use measuring cups or graduated beakers. For a large sphere you may have to use a large container. Mark the level of the water before you put the sphere into it, and then again afterwards. Next, take the sphere out, and use a measuring cup to add an equivalent amount of water to the container. Try to choose several spheres with radically different sizes. For instance, you might use a pearl, a marble, a ping pong ball, a golf ball, and a baseball. On the following graph, plot the volumes in cubic centimeters versus the diameters in centimeters.

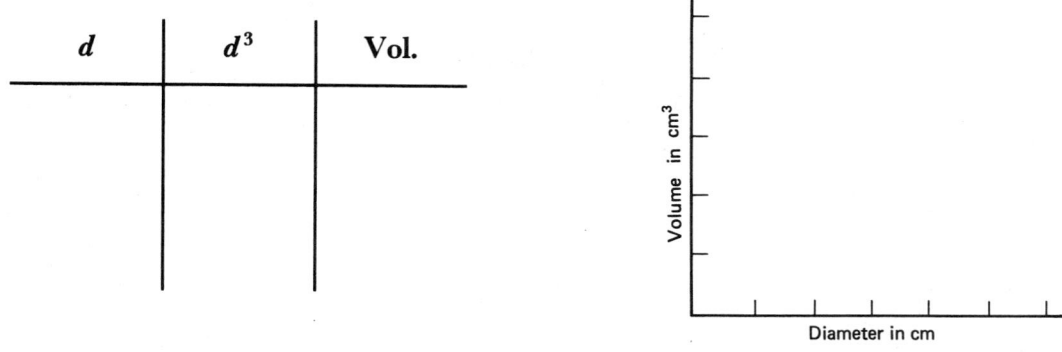

The experimental curve that you obtain will rise rapidly, but how do you know that it really shows that the volume is proportional to the diameter cubed? Here's an easy graphical test. Note that the data table includes a third column labeled $d^3$. Fill out that column from your measured data and then on the next graph plot the volume versus $d^3$, rather than the volume versus $d$. If $V \propto d^3$, then this second graph should be a straight line. Try it with your data. Don't forget to use error bars for the data regions. There are usually sizable uncertainties involved in measuring volume by water displacement. When you cube the diameter, don't forget to multiply the *relative* error by 3.

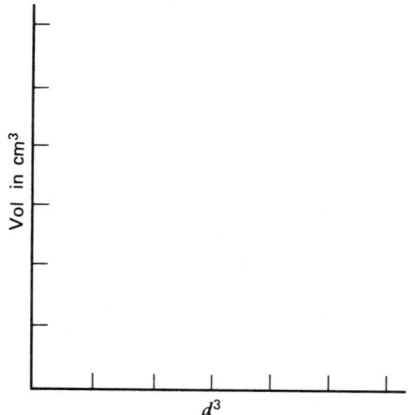

Let's take a brief look at some of the properties of the higher power functions, so that we can compare the common properties of all power functions. A complete graph of a cubic equation ($y = kx^3$) is shown in Fig. 10-29. Note that for negative values of $x$, $y$ is negative. While this feature is interesting mathematically, we will be concerned only with the positive part of the graph. Compare the rise of the cubic with that of the square.

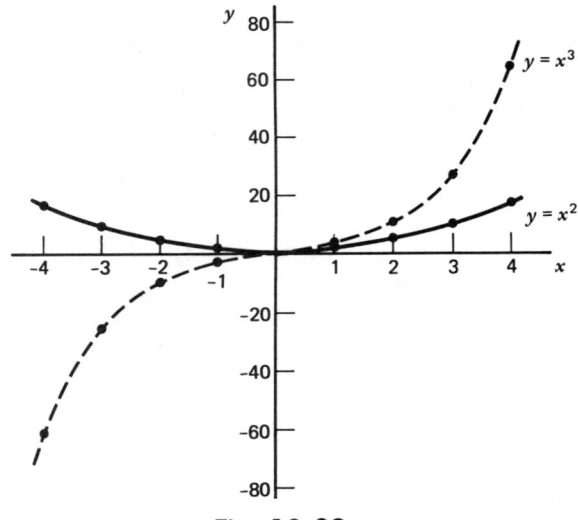

**Fig. 10-29**

The slope of a mathematical function is usually as interesting as the function itself in describing the behavior of some phenomena. We saw that the slope of a linear function is constant. The slope of the quadratic curve increases with $x$; if $y = kx^2$, then the slope steadily increases and is equal to $2kx$, a linear function. Let's take a look at the slope of the cubic curve. On the top half of the blank graph, carefully sketch the graph for positive values of $y = x^3$. Choose scale and units so that you can make the best use of the graph paper. For at least four points, including $x = 0$, draw a tangent to the curve and measure the slope. Plot these in the lower part of the graph paper.

**Question 10-15.** Do the values for the slopes of $y = x^3$ fall on a straight line?

Let's derive algebraically the slope of a tangent to the curve, $y = x^3$. The method is the same as we used when we were studying the quadratic.

$$\frac{\Delta y}{\Delta x} = \frac{y_f - y_0}{\Delta x} = \frac{(x + \Delta x)^3 - x^3}{\Delta x} = \frac{(x^3 + 3x^2 \Delta x + 3x \Delta x^2 + \Delta x^3) - x^3}{\Delta x}$$

$$= 3x^2 + 3x \Delta x + \Delta x^2$$

In the limit as $\Delta x$ goes to 0, the chord becomes the tangent with a slope of $3x^2$.

THE CUBIC: $y = kx^3$

Note the relationship between the equations for the slopes of the power functions and the power functions themselves. When $y = x^3$, the slope is equal to $3x^2$; when $y = x^2$, the slope equals $2x$; when $y = x$, the slope equals 1. In each case, the slope of a power function is also a power function, but its exponent has been decreased by 1. This is generally true of all of the power functions.

## 4. THE SQUARE ROOT: $y = kx^{1/2} = k\sqrt{x}$

**Handling the Phenomena**  A phenomenon requiring description with a square root function is easy to observe. Simply take a string and hang some small, heavy object on the end of it to make a pendulum. Hang the pendulum from a rigid support in such a way that it swings freely from a point, such as the knot (in other words, do not loop the string around a post and let the loop slither back and forth as the pendulum moves). Find the period of the pendulum as a function of the length of the string. The period of a pendulum, or of any cyclic motion, is the time taken for one complete cycle. In other words, it is the time for the pendulum bob to swing all the way over and all the way back. The length of the string pendulum should be measured to the middle of the bob, which should be small and massive.

If your string is less than a meter long, and you measure just one period, your timing uncertainty will be quite large. Even with a stopwatch, your reflexes are probably not good to better than $\pm \frac{1}{5}$ s. For a 1-s period, you would then have a 20% error. The error can be greatly reduced by the simple technique of measuring ten periods. If the total time is, for example, 10 s, and your reaction time error is still $\frac{1}{5}$ s, your percentage of error is only 2%.

> **Question 10-16.**  Could you also get a 2% measurement by measuring individual periods ten times in a row, and then taking an average?

Measure the period for at least four different lengths. Plot the data on the following graph, choosing the scales and units so that you make the best use of the whole graph paper. *Plot the data as you take it*. There is a particularly good reason for doing so, in connection with this particular exercise. If you measure the periods for lengths of 100, 90, 80, and 70 cm, you still will not know much about the functional dependence of the period on the length. If you plot those four points as you take them, the

# THE SQUARE ROOT: $y = kx^{1/2} = k\sqrt{x}$

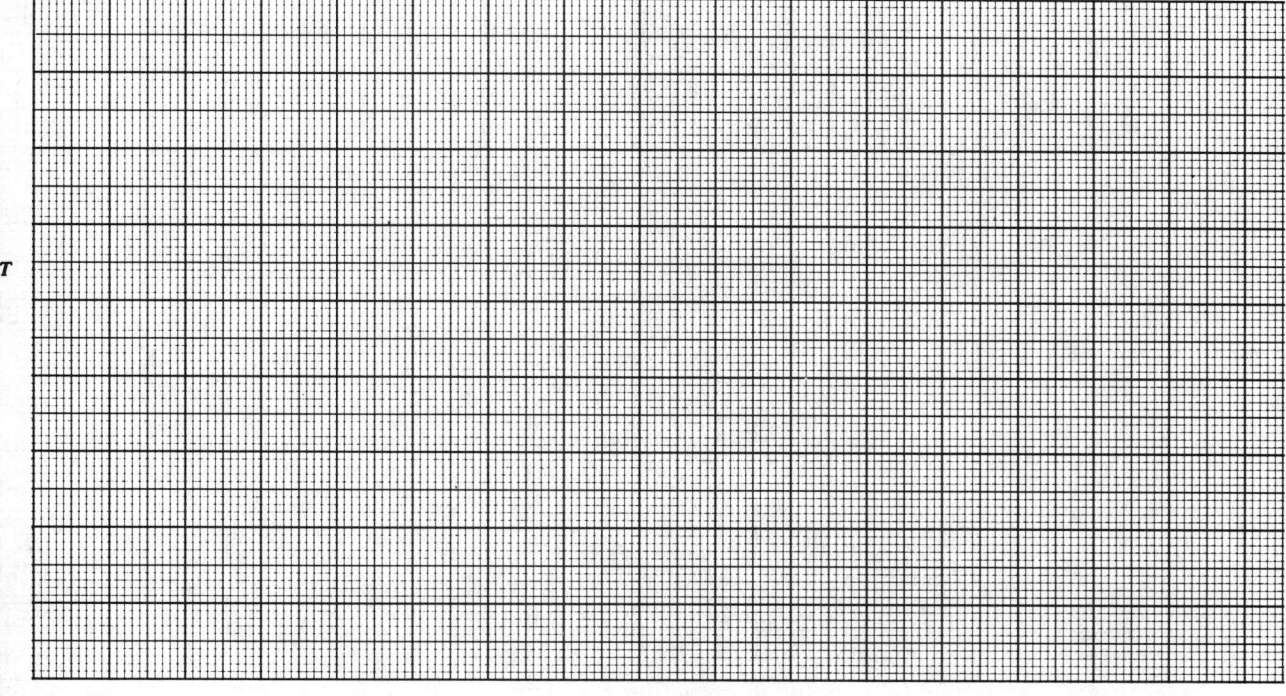

graph itself will immediately tell you that the functional dependence has not yet been determined. Instead, find the period for some very small value of the length, such as 10 cm, then for some maximum values, such as 100 cm. With those two data points on the graph, you can now shrewdly choose two more lengths that will allow you to sketch a very reasonable function curve.

Note the appearance of the function curve that you obtain. The power function curves that you have seen so far all rise up; as $x$ increases, $y$ increases, and for some of the power functions, $y$ increases very rapidly. In this case, as $x(L)$ increases, $y(T)$ also increases, but very slowly. This curve has the appearance of $y = kx^{1/2} = k\sqrt{x}$. How can we be sure, however, that such a function actually fits the data? Perhaps the true function is $T = kL^{0.4}$. To check on our assumption that nature has been kind to us, and that the function is simple, we can add a third column to our data. Find the square root of $L$ for each of the lengths you used, and list those values in column 3. Now in the next graph, plot $T$ versus $\sqrt{L}$. If the data fall on a straight line, then $T \propto \sqrt{L}$. The straightness of a line is very easy to tell. Simply sight along the line with your eyes at the same level as the line. Even small deviations from straightness become glaringly apparent. Alternatively, on the graph paper, simply see if you can draw a straight line through the data points with a ruler.

In the top half of the next graph, draw the curve for $y = x^{1/2}$, choosing units and scales so that you make the best use of the graph paper. Now on four different points, including $x = 0$, draw tangents and find the slopes. Plot these slopes in the bottom half of the graph.

**Question 10-17.** How does this graph, showing the slope function for $y = x^{1/2}$, differ from the slope functions that we have seen for the other power functions?

The algebraic derivation of the formula for the slope of the square root function requires the use of series expansions of functions. The result of the derivation is that the slope function of $y = x^{1/2}$ is equal to $\frac{1}{2}x^{-1/2}$. Note how our rule for the slopes of power functions still holds. The exponent of the slope function is smaller than the exponent of the power function itself by 1 (in this case, from $+\frac{1}{2}$ to $-\frac{1}{2}$).

THE SQUARE ROOT: $y = kx^{1/2} = k\sqrt{x}$

97

$y$

Slope of $y$ ─────────────────────────────────── $x$

$x$

## 5. POWER FUNCTIONS WITH NEGATIVE EXPONENTS

At constant temperature, the pressure of a gas is inversely proportional to its volume: $P = k/V = kV^{-1}$. The potential difference between a point with an electric charge and another point some distance away is inversely proportional to the distance between them: $V = kq/r = kqr^{-1}$. The electric field (force per unit charge) at a distance from a point charge is inversely proportional to the square of the distance: $E = k/r^2 = kr^{-2}$. Gravitational fields produced by spheres also depend on the inverse square of the distance from the center of the sphere. Newton's law for the gravitational attraction between two distant spheres is $F = G\dfrac{m_1 m_2}{r^2}$.

In Fig. 10-30, we have plotted the functions $y = kx^{-1}$ and $y = kx^{-2}$. Let's take a closer look at these inverse, or negative power, functions. In the first place, note that for small values of $x$, $y$ is large, and vice versa. The first of the equations, $y = kx^{-1}$, has a formal name: it is a hyperbola.

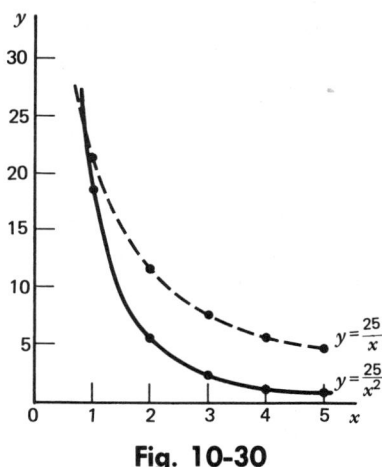

**Fig. 10-30**

The inverse square function describes any influence, such as an electric or gravitational field, that spreads out equally in all directions from a point, or spherical, source. Other such influences that follow such a law are light intensity (from a small light bulb, for instance) or sound intensity (from a small loudspeaker).

Let's see whether the slope of a negative power function follows the same rule that we deduced for the slopes of positive power functions. Draw tangents at four different points to the graph $y = x^{-1}$. Find the slopes of these tangents, and plot them on the second graph.

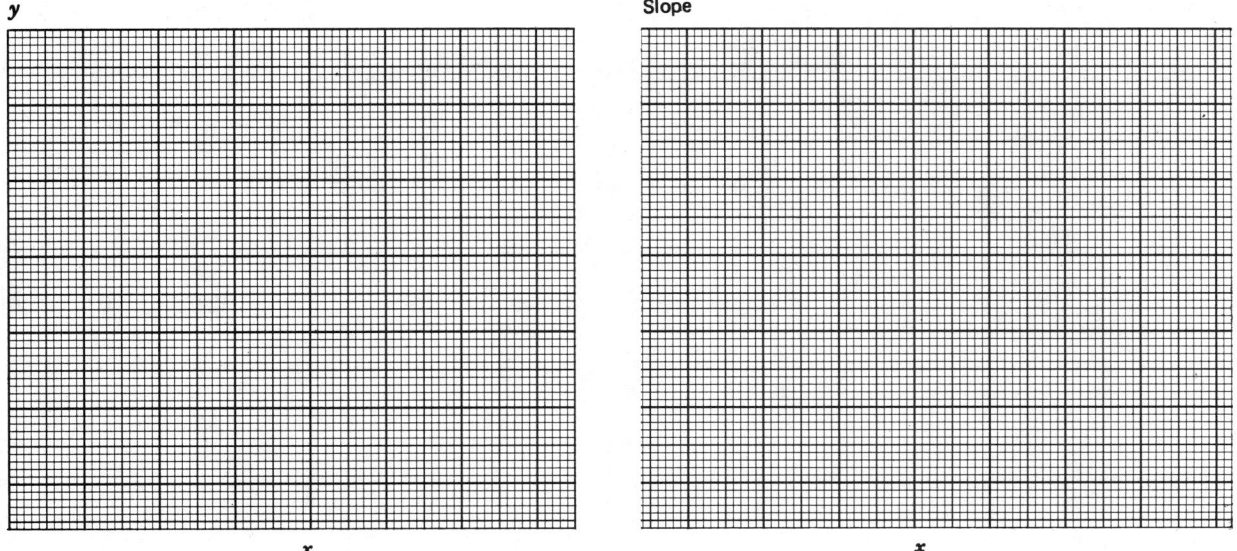

**Question 10-18.** If the function describing the slopes of $y = x^{-1}$ is also a power function, about what value must the exponent have?

We can derive algebraically the function that describes the slope of $y = x^{-1}$, using the same technique that we used before:

$$\frac{\Delta y}{\Delta x} = \frac{1}{\Delta x}\left(\frac{1}{x + \Delta x} - \frac{1}{x}\right) = \frac{1}{\Delta x}\left(\frac{x - (x + \Delta x)}{(x + \Delta x)x}\right)$$

$$= -\frac{1}{x^2 + x\Delta x}$$

As $\Delta x$ goes to 0, $\Delta y/\Delta x$ goes to $-x^{-2}$.

**Question 10-19.** What is the rule that we discovered for the slope functions of power functions? Does the slope of $y = x^{-1}$ follow this rule?

**Handling the Phenomena** Why should an influence spreading out from a point source decrease as the square of the radius? In the first place, the mathematical model is valid only if the influence spreads evenly in all directions and does not decay and is not absorbed. If these conditions are fulfilled, then the same amount of "influence" passes through each successive, concentric sphere that you put around the source. Since the areas of the concentric spheres increase with radius, the amount of influence *per area* must decrease with increasing radius.

Get two spheres of different sizes, such that one has two or three times the radius of the other (perhaps a ping pong ball and a hard baseball). Draw an equator and a point for a north pole on each sphere. Divide the equators by eye approximately into quarters. On each sphere, divide one of these quarters again by eye approximately into four segments. If these were parts of the equator of the earth, each segment would have an angular width of $22\frac{1}{2}°$. Now draw longitudinal lines from the north pole to the ends of one of these segments on each sphere. You now have two spherical triangles. Because the apex angles are small, you can get a good approximation to their areas by using the formula for the area of flat triangles: $A = \frac{1}{2}$(base)(height). Measure the areas of both of these triangles using a flexible plastic ruler or string to measure the bases and heights. What is the ratio of the diameters of the two spheres? _____. What is the ratio of the triangular areas of the two spheres? _____.

The surface area of a sphere is $A_{\text{sphere}} = 4\pi r^2 = \pi d^2$. You measured only a segment of a sphere, but if the whole sphere has a surface area proportional to the radius squared, then any angular segment of the sphere should also have an area proportional to the square of the radius. When an influence, such as light or sound, spreads out from the center of a sphere, the amount of influence passing through the whole spherical surface of any radius remains the same. However, the amount of influence *per unit area* decreases as the square of the radius, because the area of the sphere is *increasing* as the square of the radius.

## 6. COMPARISON OF THE VARIOUS POWER FUNCTIONS

In Fig. 10-31, we have drawn a number of the power functions to illustrate their similarities and differences. The larger the exponent in the power function, the faster it rises. However, note what happens in the region from $x = 0$ to $x = 1$.

**Question 10-20.** Why does the curve for $y = x^3$ lie underneath the curve for $y = x$ from $x = 0$ to $x = 1$?

If you are plotting data from some experiment and the curve rises faster than a straight line, then there is a possibility that it can be fitted with a power function whose exponent is greater than 1. If you are fitting data from some experiment and the curve rises continuously, but less steeply than a straight line, then a power function may satisfy the data if the exponent is greater than 0 but less than 1

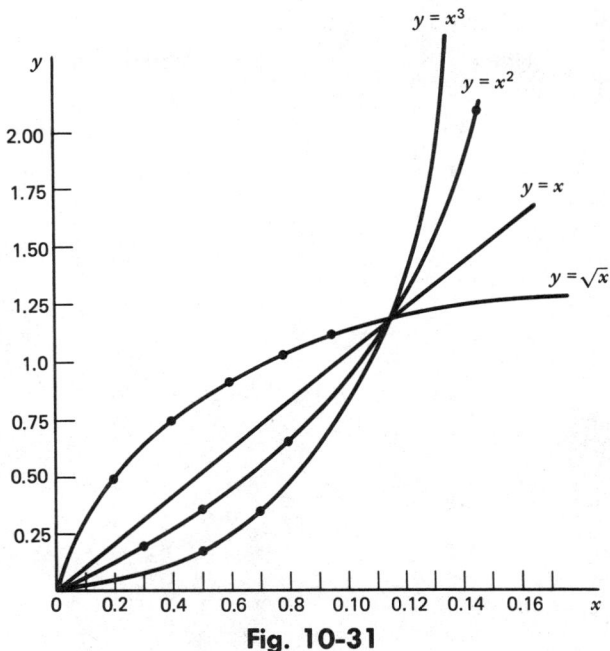

**Fig. 10-31**

(for instance, $\frac{1}{2}$ or the square root function). If you are doing an experiment and find that one variable decreases while the other increases, then it is possible that a power function with negative exponent may satisfy the data. On the other hand, we will see another function that also has this same characteristic.

## Answers to Questions

10-1. As shown on the graph, the savings at 52 weeks is $520. At 26 weeks, it is $260. Doubling the time, from 26 weeks to 52, has doubled the money. Proportionality holds throughout.

10-2. If you plot your age versus the year date in the century, you get a straight line. The two variables are linearly related. However, a person's age can be proportional to the year date in the century only if the birth date is 1900. In that case there is no intercept.

10-3. These are similar right triangles. The measure ratios all have the same value.

10-4. $y = mx + b$

$0 = mx + b$

$x = -b/m$

10-5. $60 \, \frac{\text{miles}}{\text{h}} \left( \frac{1 \text{ h}}{3600 \text{ s}} \right) \left( \frac{1609 \text{ m}}{1 \text{ mile}} \right) \approx 27 \, \frac{\text{m}}{\text{s}}$

$1 \, \frac{\text{m}}{\text{s}} \left( \frac{1 \text{ mile}}{1609 \text{ m}} \right) \left( \frac{3600 \text{ s}}{1 \text{ h}} \right) = 2.2 \, \frac{\text{miles}}{\text{h}}$

10-6. $x = (120 \text{ miles}) + (60 \text{ mi/h})(3 \text{ h}) = 300 \text{ miles}$

10-7. $\$ = -(\$1000/\text{yr})t + \$10,000$

| $t$ | $\$$ |
|---|---|
| 0 | 10,000 |
| 5 | 5,000 |
| 10 | 0 |
| $-1$ | 11,000? |
| 12 | $-2,000$? |

For $t = -1$ yr and $t = 12$ yr, we have carried the application of the model too far. The domain of the model is only from $t = 0$ to $t = 10$.

10-8. To find your weight in newtons, measure your weight in pounds and multiply by 4.45. Evidently, it is much easier with dieting to lose a newton than a pound.

10-9. The formula for $y(t)$ for an object falling freely is $y = \frac{1}{2}(9.8)t^2$, where $y$ is in meters and $t$ is in seconds. Since the slope of $y = kx^2$ is equal to $2kx$, the slope of $y = \frac{1}{2}(9.8)t^2$ is 9.8 ($k = \frac{1}{2}9.8$). In this case the slope is $\Delta y/\Delta t$ which is the vertical speed of fall: $v = 9.8t$. Check to see if your measured values of the slopes agree with this formula.

10-10. Acceleration is the time rate of change of velocity: $\Delta v/\Delta t$. The units of $g$ are (m/s)/s, which is correct. In the equation $y = \frac{1}{2}gt^2$, the dimensions on the left are $L$. On the right they are $\left(\dfrac{L}{T^2}\right)T^2$, which also equals $L$.

10-11. $\dfrac{\Delta y}{\Delta t} = \dfrac{y_f - y_0}{\Delta t} = \dfrac{v_0(t + \Delta t) + \frac{1}{2}g(t + \Delta t)^2 - v_0 t - \frac{1}{2}gt^2}{\Delta t}$

$= \dfrac{v_0 \Delta t + gt \Delta t + \frac{1}{2}g \Delta t^2}{\Delta t} = v_0 + gt + \frac{1}{2}g\Delta t$

$\rightarrow v_0 + gt$

The final speed is the sum of the original speed $v_0$ plus the speed $gt$ due to the acceleration.

10-12. The "area" under the curve of a graph represents the product of the ordinate and the abscissa. Each of these has units and dimensions that are not usually those of length. In this case, the area represents the product of velocity and time. (m/s) × s → m. $(L/T) \times T \rightarrow L$. The graph area represents the distance traveled.

10-13. Acceleration is a vector quantity. If the acceleration is negative (to the left) and the original velocity is positive (to the right), the speed is reduced. When the velocity becomes negative also, the negative acceleration increases the speed to the left.

# COMPARISON OF THE VARIOUS POWER FUNCTIONS

10-14. $V = \frac{4}{3}\pi\left(\frac{d}{2}\right)^3 = \frac{4}{3}\frac{1}{8}\pi d^3 = \frac{1}{6}\pi d^3 \approx \frac{1}{2}d^3$

Note that the volume of a sphere is about one-half the volume of the cube ($d^3$) into which the sphere will just fit.

10-15. They'd better not.

10-16. No. The averaging method of reducing error works only if the errors are random, and even then the method is complicated. If you repeat such a measurement ten times, you will probably repeat the same error of observation ten times.

10-17. For all the other power functions, $y = kx^n$, the slope function (the derivative) is equal to $nkx^{n-1}$. For $kx^{1/2}$, our rule would lead to a slope function of $\frac{1}{2}kx^{-1/2} = \frac{k}{2\sqrt{x}}$. This is a reciprocal function with large values for small $x$ and small values for large $x$.

10-18. The slope is always negative, and so the factor must be negative. The slope starts with large values for small $x$ and goes to small values for large $x$. Therefore, the exponent must be negative.

10-19. Yes. If $y = kx^n$, the slope function (derivative) is $nkx^{n-1}$. So the slope function of $kx^{-1}$ should be $-kx^{-2}$.

10-20. For $x < 1$, $x^2 < x$. If $x = \frac{1}{2}$, $x^2 = \frac{1}{4}$, and $x^3 = \frac{1}{8}$.

# CHAPTER 11
# FORCE AND NEWTON'S SECOND LAW

If you pull on something, you may stretch it. For certain types of springs, as we saw on page 76, the stretch is proportional to the pull: $F = kx$. In this formula (known as Hooke's law), $x$ is the stretch, measured in meters; $F$ is the force, or pull, measured in newtons; and $k$ is the proportionality constant with units of newtons per meter (N/m). If you double the force, you double the stretch.

**Question 11-1.** A typical laboratory spring is 10 cm long and has a spring constant $k = 200$ N/m. If you pull on the spring with a force of 200 N, what is the new length?

## THE FORCE OF WEIGHT

The "weight" of an object is the gravitational force between the object and the earth. The earth pulls on the object; the object pulls with the same force on the earth. The gravitational attraction depends on the distance between the earth and the object. If the object were in space, its "weight" would be less.

The formula for the gravitational attraction between two objects was given by Newton:

$$F = G\frac{mM}{r^2}$$

For our definition of weight, $m$ is the mass of the object (perhaps you) in kilograms; $M$ is the mass of the earth in kilograms ($6 \times 10^{24}$ kg); $r$ is the distance between the object and the center of the earth in meters (if the object is on the surface of the earth, we have $r = 6.4 \times 10^6$ m); and $G$ is the universal gravitational constant ($6.7 \times 10^{-11}$ Nm$^2$/kg$^2$).

**Question 11-2.** Find your weight in newtons if your mass is 70 kg (154 lb).

On the surface of the earth your weight is equal to $W = F = GmM/r^2 = m(GM/r^2)$. The value of the quantity in parentheses is $(6.7 \times 10^{-11})(6 \times 10^{24})/(6.4 \times 10^6)^2 = 9.8$ N/kg. This quantity is called the gravitational field on the surface of the earth and is given the special symbol $g$. (Note that $G$ is a universal constant, used in Newton's formula; $g$ applies only to the surface of the earth.) Another name for $g$ is "the acceleration due to gravity." That is because if you drop any object, regardless of mass, it will fall with the constant acceleration of $g$ near the surface of the earth (if it is in a vacuum).

**Question 11-3.** What is your weight in newtons if your mass is 70 kg?

For many calculations, it is sufficiently accurate to let $g \approx 10$ N/kg or 10 m/s$^2$.

## STRENGTH TESTS

If possible, use a Newton spring-scale to measure the following forces. When you can't use a spring scale, calculate or estimate the force (1.00 lb = 4.45 N).

1. The weight of this book
2. The force needed to slide this book along a smooth table
3. The weight of a ball-point pen
4. The force that you can exert by pressing down your index finger without moving the rest of your hand
5. The maximum force that you can exert by pulling in opposite directions with your two hands
6. The weight of a bicycle
7. The weight of any particular car
8. The weight of the earth in *your* gravitational field

## PRESSURE

You may be strong, up to a point. But can you easily resist a force of 1 N on the tip of one finger? That depends on how the force is applied! In both cases shown in Fig. 11-1, the weight on the finger is $mg = (0.1 \text{ kg})(9.8 \text{ N/kg}) = 1$ N.

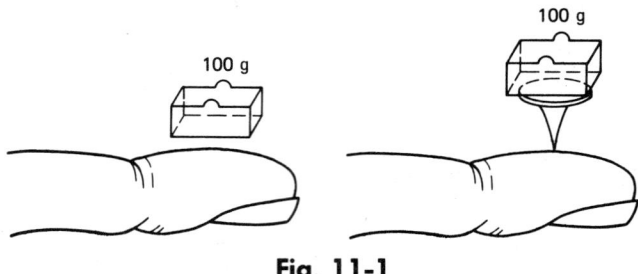

**Fig. 11-1**

*Pressure is the force per unit area.*

$$P = \frac{F}{A}$$

In the SI system, pressure is measured in newtons per square meter. The unit has a special name, the *pascal*:

$$1 \text{ N/m}^2 = 1 \text{ Pa}$$

If the 100 g object in the diagram above has a cross section area of 1 cm², then the pressure it exerts on the finger is $P = (1 \text{ N})/(1 \text{ cm}^2) = (1 \text{ N})/(1 \times 10^{-4} \text{ m}^2) = 1 \times 10^4$ Pa. If the object is resting on the thumbtack, the pressure on your skin is more likely $P = (1 \text{ N})/(0.01 \text{ mm}^2) = 1 \text{ N}/(1 \times 10^{-8} \text{ m}^2) = 1 \times 10^8$ Pa.

Standard atmospheric pressure at sea level is $1.0 \times 10^5$ Pa. We don't feel the pressure because it is on all sides of us, including inside. If we create a vacuum, we remove the atmospheric pressure.

**Question 11-4.** How high can a good suction pump "lift" water in a well? The density of water is $1.0 \times 10^3$ kg/m³.

Note that in this problem, the cross sectional area of the well pipe was unimportant. (A pipe with a cross section of 1 m² would be tremendous!) If you reduce the area by a factor of 1000, the upward force is less by a factor of 1000. However the weight of water to be supported is also less by a factor of 1000. The height remains the same.

During a severe wind storm, air pressure might fluctuate by 1% in a sudden gust. If the air pressure in a closed house remains constant during the gust, what is the total force on a window that has an area of 1 m²?

$$1\% \text{ of } 1 \times 10^5 \text{ N/m}^2 = 1 \times 10^3 \text{ N/m}^2 = \text{change of pressure}$$

$$\text{extra force} = \text{area} \times \text{extra pressure} = (1 \text{ m}^2)(1 \times 10^3 \text{ N/m}^2) = 1 \times 10^3 \text{ N}$$

Since $1 \times 10^4$ N ≈ 1 ton, the sudden unbalanced force on the window would be about 0.1 ton.

# THE OTHER PROPERTY OF FORCE

Force distorts objects. Indeed, that is the way we have defined forces so far. They distort (stretch) the springs in spring scales. *If the net force acting on an object is zero*, the object may be distorted but we expect it to be in equilibrium.

$$\text{if } F_{net} = 0, \text{ the object is in equilibrium}$$

Equilibrium does not necessarily mean motionless! The object might be standing still or might be moving with constant velocity. For instance, if a car is moving at constant velocity, the forward force provided by the road friction (in response to the pressure of the drive wheels) must just balance the backward force of wind and tire friction of the free wheels. If an object is moving with constant velocity with respect to us (and the velocity might be zero), then the net force acting on it (with respect to us) is zero.

If the net force on an object is *not* zero, the velocity of the object is not constant. Instead, the object is accelerating. *Newton's second law of motion is*

$$\boxed{F = ma}$$

Acceleration is proportional to the net applied force. The proportionality constant is the mass of the object.

**Question 11-5.** A pull of 10 N is applied to one end of a long coil spring (like a slinky) that is lying on the ground. The mass of the spring is 0.5 kg. What is the result?

Suppose you apply 1.0 N to an air track glider which can move with almost zero friction. The force is parallel to the track and the mass of the glider is 100 g. What is the acceleration?

$$F = ma \qquad 1.0 \text{ N} = (0.1 \text{ kg})a$$
$$a = 10 \text{ m/s}^2$$

How long will it take the glider, starting from rest, to go 2.0 m?

$$x = \tfrac{1}{2}at^2 \qquad 2.0 \text{ m} = \tfrac{1}{2}(10 \text{ m/s}^2)t^2$$
$$t^2 = 0.40 \qquad t = 0.63 \text{ s}$$

**Question 11-6.** If you drop a ball with a mass of 0.5 kg, what is its downward acceleration?

Suppose you exert a force of 100 N on a large box that has a mass of 100 kg and is resting on the floor. What happens depends on the direction of your force! If you press straight down, you will compress the box slightly, but you won't accelerate it. If you pull horizontally, the box will probably move, but there is friction. The friction acts as a constant force (independent of velocity) in the direction opposite to the velocity. In this case, suppose the friction force is 30 N. Then the *net* force is 100 N − 30 N = 70 N. The acceleration of the box is found from $F = ma$:

$$(100 \text{ N} - 30 \text{ N}) = (100 \text{ kg})a$$

$$a = 0.7 \text{ m/s}^2$$

## Answers to Questions

11-1. If we used the formula blindly, the stretch would be 1 m and the spring would then be 110 cm long. However, few springs can stretch to eleven times their original length. Here is an example of how formulas and rules are only models of physical phenomena. Models apply only within limited ranges and conditions. In this case, Hooke's law would probably not apply, long before the stretch got to be 1 m.

11-2. Your weight is $W = F = GmM/r^2$

$$= (6.7 \times 10^{-11})(70 \text{ kg})(6 \times 10^{24} \text{ kg})/(6.4 \times 10^6 \text{ m})^2 = 690 \text{ N}$$

11-3. Instead of the lengthy calculation of the preceding question, you can find your weight as $W = mg = (70 \text{ kg})(9.8 \text{ N/kg}) = 690 \text{ N}$

11-4. Of course, the pump doesn't "lift" the water. The pump removes the air in a tube above the water. The water in the well is under atmospheric pressure and is shoved up the tube, as shown in Fig. 11-2. If the tube had a cross sectional area of 1 m², the upward *force* on the water in the tube at the level of the well water would be $F = PA = (1 \times 10^5 \text{ N/m}^2)(1 \text{ m}^2) = 1 \times 10^5 \text{ N}$. That force must support the weight of the water in the tube:

$$(1.0 \times 10^3 \text{ kg/m}^3)(9.8 \text{ N/kg})(\text{height}) = \text{weight in tube} = 1.0 \times 10^5 \text{ N}$$

$$\text{height} = 10.2 \text{ m} = 33.5 \text{ ft}$$

Most household well pumps can't raise water more than about 25 ft.

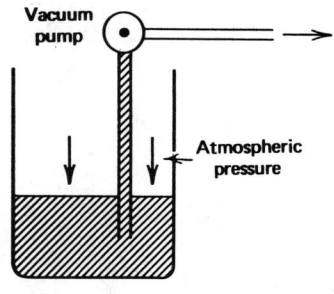

**Fig. 11-2**

11-5. The applied force of 10 N is not the only force acting on the spring. Gravity is pulling the spring downward and there is a friction force if the spring is pulled along the ground. What happens depends on the direction of the applied force and the amount of friction. In most cases, the applied force will cause two effects. It will *stretch* the spring and it will *accelerate* it. The amount of the stretch will be proportional to the resisting force (combination of weight and friction). The amount of acceleration will be proportional to the *net* force—the difference between the applied force and the resisting force.

11-6. The weight of the ball is

$$W = mg = (0.5 \text{ kg})(10 \text{ N/kg}) = 5 \text{ N}$$

The weight is the downward force:

$$F = ma \quad (5 \text{ N}) = (0.5 \text{ kg})a$$

$$a = 10 \text{ m/s}^2$$

But note! The effect does not depend on the mass of the ball. $W = mg = F = ma$. Since $mg = ma$, $a = g$, regardless of the mass. The mass cancels out. (In your physics course, you will probably learn that this situation is surprising, and not at all obvious. The "gravitational mass" does equal the "inertial mass," but we still aren't sure why this is true.) Because of this fact, the "gravitational field" $g = 9.8$ N/kg is equal to the "acceleration due to gravity" $g = 9.8$ m/s$^2$.

## PROBLEMS

1. What is the pressure at the bottom of a swimming pool 3 m deep?

2. If you have a tug of war with each side pulling with a force of exactly 1000 N, as shown in Fig. 11-3, what will be the reading of a newton spring scale inserted into the rope?

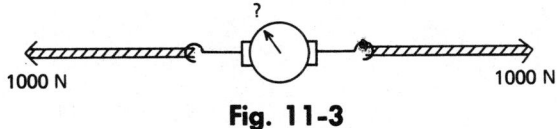

Fig. 11-3

3. For a particular spring, $k = 1000$ N/m. What force is required to stretch the spring 1 m?

4. To compress two pieces of wood that are being glued, a force of $1 \times 10^4$ N is exerted on a surface 2 cm $\times$ 5 cm. What is the pressure?

5. What is the total force on the side of a swimming pool that is 10 m long and is filled with water to a height of 2 m?

6. A force of 30 N is exerted straight up on a box that has a mass of 2 kg. What is the acceleration?

7. A horizontal force of 100 N is exerted on a cart which has a mass of 20 kg. There is a resisting friction force of 60 N. Starting from rest, how long will it take the cart to go 100 m? How fast will it be going then?

8. Suppose you want to push a pile driver straight down with an acceleration of 15 m/s$^2$. Its mass is 10 kg. What force must you exert?

# CHAPTER 12
# VECTOR ADDITION

Stand up. Take two steps forward, then take two more steps forward. How far have you gone? This must seem like a silly question. 2 + 2 = 4. Now, start again, and take two steps forward. Turn 90° and take two more steps. How far have you gone? The answer is not so obvious. You have, of course, traveled four steps, but you are not four steps from your starting place. In terms of *displacement*, or actual distance from the origin, 2 + 2 ≠ 4.

"Displacement" is a technical term. To describe displacement, you must provide both the *magnitude* of the distance traveled and the *direction* with respect to some coordinate system. A displacement is the prototype of *vector* quantities.

Not every variable that has magnitude and direction is necessarily a vector. There is no point, for instance, in describing a mountain in terms of a vector, and yet mountains clearly have magnitude and direction (up). On the other hand, as we shall see, some quantities which at first glance do not appear to have direction can be represented as vectors. To be represented as a vector, a quantity must have magnitude, direction *and* must combine with other vectors in a particular way characterized by the behavior of displacements. Variables that have only magnitude are called *scalars*. Time, for instance, is a scalar quantity. Of course, you can define plus time and minus time, having chosen $t_0$, but time can not be assigned directions, north, south, east, or west.

**Question 12-1.** Here are a few variables. From what you know now, categorize them as being either vectors or scalars: displacement, velocity, acceleration, mass, temperature, force, volume, density, area.

Vectors are frequently represented in diagrams by arrows. These arrows are simply scale models of displacements, with the arrowhead indicating the direction of motion. In Fig. 12-1, we have drawn

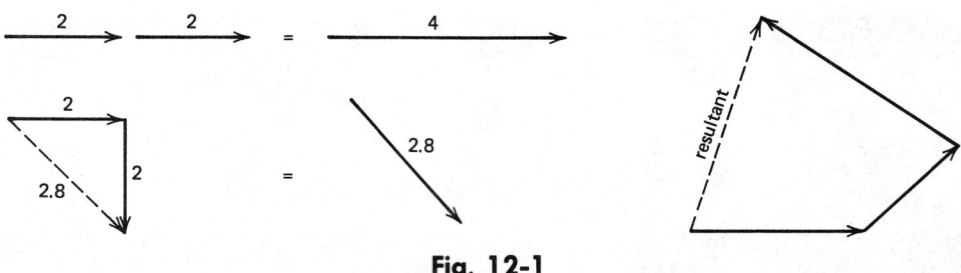

**Fig. 12-1**

several combinations of displacements, including the two with which we started out this section. When we combine vectors like this, we say that we are "adding" them. As you can see, vector addition is very different from ordinary addition, since 2 + 2 can equal anything from 0 to 4. Furthermore, the *resultant* not only has magnitude but also has direction.

While it is useful to sketch diagrams of vector combinations, we should use a different analysis for actual calculations (no one wants to carry a drafting board all the time). Usually, the easiest way to deal with vectors is to break them down into their *components*. This is called *resolving* the vector. In Fig. 12-2, we illustrate this technique with a number of vectors, leaving some of them blank for you to complete.

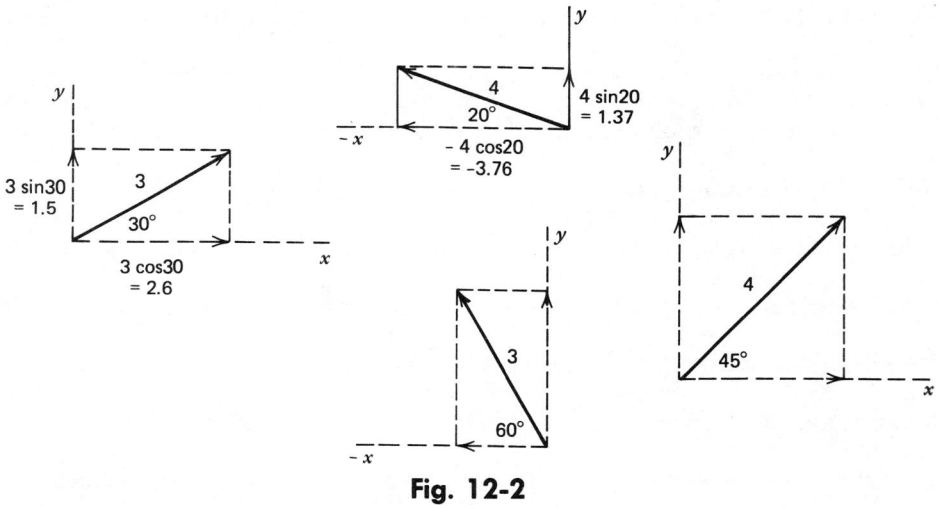

**Fig. 12-2**

In taking components of vectors, we are only using right angle trigonometry. Usually, we are used to such definitions as $\sin\theta = \dfrac{\text{opp}}{\text{hyp}}$. In taking components of vectors, we use the same expression in the form: opp = hyp $\sin\theta$ and adj = hyp $\cos\theta$. Of course, there's nothing new in these forms, but since you will use them so often, it is best to get used to them rather than having to go back each time to the original definition of $\sin\theta$ and $\cos\theta$.

Let's take a journey. We will travel 6 m east, 4 m at 30° north of east (N of E), 2 m west and 6 m SW. We have sketched this series of displacements in Fig. 12-3. Note that the displacement arrows are linked head to toe. The abbreviation SW is the standard technical way of writing southwest, and, in particular, 45° south of west, or west of south. Similarly, NE means along the 45° line between north and east. Now let's add, or combine, these displacements. The question is, having taken the journey, where are you with respect to your starting point? Your *resultant displacement* is a vector with magnitude and direction. It is shown in the sketch as a dotted line. Even though we are going to solve the problem analytically by finding the components of the individual vectors, it is useful to have a sketch roughly drawn to scale in order to check the plausibility of our final answer.

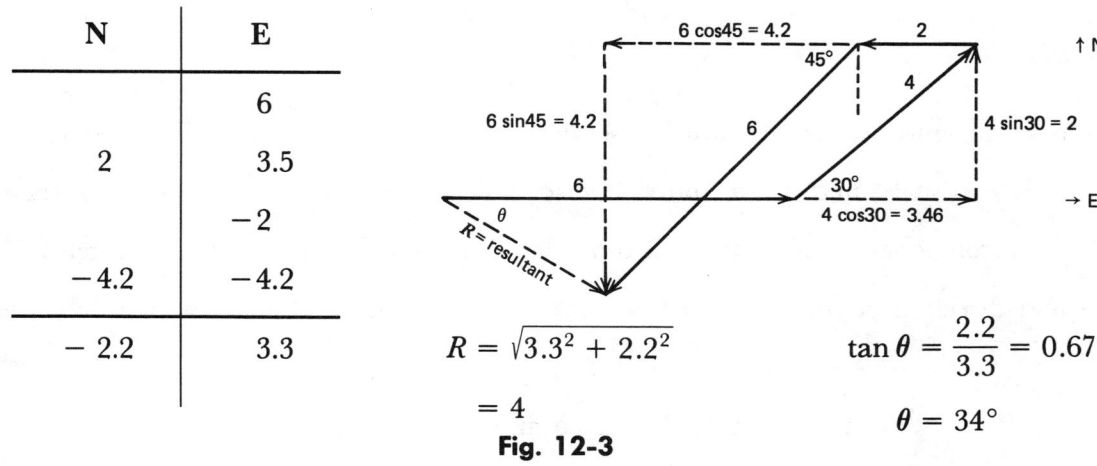

| N | E |
|---|---|
|   | 6 |
| 2 | 3.5 |
|   | −2 |
| −4.2 | −4.2 |
| −2.2 | 3.3 |

$R = \sqrt{3.3^2 + 2.2^2}$

$= 4$

$\tan \theta = \dfrac{2.2}{3.3} = 0.67$

$\theta = 34°$

**Fig. 12-3**

On the diagram, each vector has been broken into perpendicular components along the NESW directions. These components are then labeled in terms of the length of the displacement and the appropriate trig function. At this point, we have reduced the problem to two very simple ones. We can add up all the components along the east–west direction as if they were ordinary scalars, and then do the same thing for the N–S components. A component in the W direction should be labeled as a negative component in the E direction. Add up the N and E components separately, as shown in the tally columns with the diagram. What we end up with are the components of the resultant displacement. From these two components, we can compose the resultant using the Pythagorean theorem. The orientation of the resultant can be given in terms of $\theta$ where $\tan \theta$ equals the ratio of the two components. Take a close look at our original sketch and see if the magnitude and direction of the calculated resultant agree approximately with the appearance of the resultant on the sketch.

**Question 12-2.** What happens if you reverse the sequence of displacement? Start out with the move of 6 m to the SW and end up with the move, 6 m to the E. Draw the sketch of this sequence, and see if you can find a general argument why **A** + **B** = **B** + **A**. (Note that we use boldface type for vector quantities. When writing by hand, we indicate vectors by drawing an arrow over the symbol. For instance, $\vec{A} + \vec{B} = \vec{B} + \vec{A}$.)

In the next student work space a different sequence of displacements is described. Sketch the given sequence roughly to scale, and use a dotted line to indicate the resultant. Then break each of the vectors up into components, label these components and add the two sets separately. Find the magnitude and direction of the resultant and make sure that the results agree plausibly with your sketch.

3 m  N, 6 m 30° S of E, 5 m W, 4 m 45° N of W

To be a vector, any other quantity must follow the addition (combination) rules that we have just described for displacements. In some physics problems, the vectors will exist in three dimensions and so must be broken up into three components each. We have used perpendicular components of north and east. In more general terms, the components might be along the $x$, $y$, and $z$ axes. Fig. 12-4 shows the

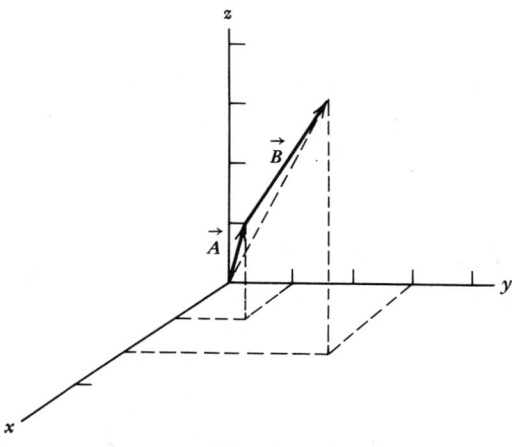

**Fig. 12-4**

addition of two vectors in three dimensional space. The process of resolving each vector into components is the same as we have already done, except that in this case there are three components. The magnitude of the resultant sum is $|A + B| = \sqrt{(A_x + B_x)^2 + (A_y + B_y)^2 + (A_z + B_z)^2}$.

Although usually it is most convenient to resolve vectors into the perpendicular directions $x$, $y$, $z$, sometimes, as we shall see, other perpendicular directions are more convenient. In any case, the principle is the same. Break the vector down into components which can be separately added as scalars and then recombined to form the resultant vector.

If you can pull on a rope tied to a cart with a force of 100 N (newtons), and somebody else also has a rope tied to the cart and is pulling with a force of 100 N, what is the resultant force on the cart? Once again, we are faced with a vector problem—or at least we *may* be faced with a vector problem. Forces certainly have magnitude and can be exerted in any direction. Whether or not they follow the combination laws for vectors depends on experimental demonstration. It seems plausible that forces might combine like displacements. If you and your friend are pulling in opposite directions on the cart, the net force on the cart will be zero. If you are pulling in the same direction, it seems reasonable to assume that the resultant force is 200 N.

**Handling the Phenomena** You will need three calibrated spring scales. Lacking those, do the experiment qualitatively, with three rubber bands. Arrange these force measuring devices in the several different combinations shown in Fig. 12-5. In the case where the forces are all acting in line, you can check to see whether or not forces add linearly; that is to say, is it the case that 2 N + 2 N = 4 N if they are acting in the same direction? Of course, you would assume that to be the case, but not everything in this world combines in a linear fashion. For instance, if you double the force on a rubber band, you will usually not double the stretch. For the other cases shown, break each force vector into

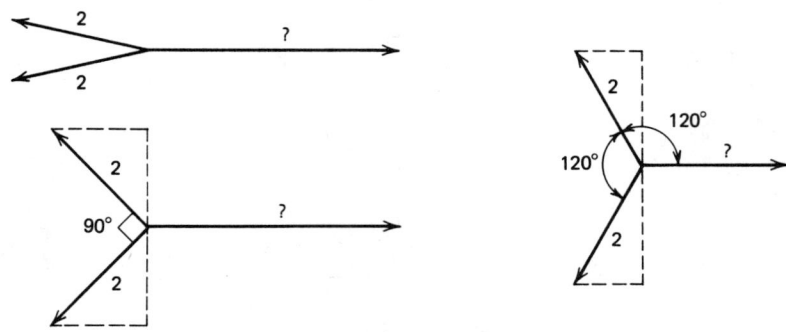

**Fig. 12-5**

$x$, $y$ components. If the forces are in balance, all the $x$ components must equal zero and all the $y$ components must equal zero.

To get the missing numbers to put into the diagram, arrange the spring scales in the angular positions shown and read the forces that they exert. Then do the arithmetic to see whether the theoretical results agree with the experimental.

## THE EFFECTIVE COMPONENTS OF FORCE

An unbalanced force acting on an object will accelerate it according to Newton's second law. The acceleration is in the direction of the force. Frequently, however, only one component of the applied vector force is effective in accelerating the object. The other, perpendicular, component may do nothing or may change the friction resisting the motion.

Suppose we pull a sled with a rope at an angle of 30° with the horizontal, as shown in Fig. 12-6.

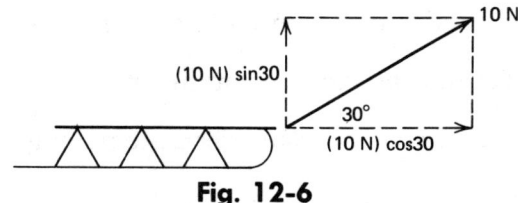

**Fig. 12-6**

The effective pulling component is horizontal and equal to $(10 \text{ N})\cos 30° = 8.7$ N. The vertical component of 5 N tends to lift the front of the sled. If we assume that friction is negligible, and if the sled and rider have a combined mass of 30 kg, the acceleration is

$$a = F_{\text{eff}}/m = (8.7 \text{ N})/(30 \text{ kg}) = 0.29 \text{ m/s}^2$$

In the case of the sled, the force vector was most conveniently resolved into vertical and horizontal components. In Fig. 12-7 we show a situation where the original vector is already vertical, but it is more convenient to find perpendicular components at other angles. Suppose a block can slide without friction down an inclined plane.

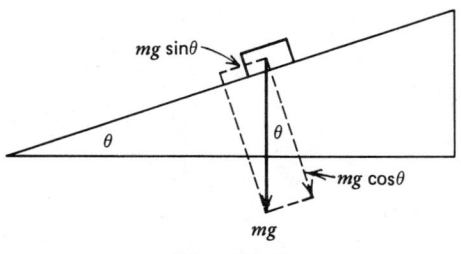

**Fig. 12-7**

The weight of the block $mg$ is vertical. The useful components are parallel and perpendicular to the plane. The parallel component ($mg \sin \theta$) accelerates the block down the plane. The perpendicular component ($mg \cos \theta$) presses the block into the plane. If there were friction this perpendicular component would be important. Without friction, the acceleration of the block is given by

$$F_{\text{eff}} = ma$$
$$a = (mg \sin \theta)/m = g \sin \theta$$

**Question 12-3.** Does our formula agree with everyday experience? What happens if $\theta = 0°$? What happens if $\theta = 90°$?

Figure 12-8 shows a block on an inclined plane. Draw the force vector and components and label them. Find the acceleration. How long would it take the block, starting from rest, to slide 1.0 m?

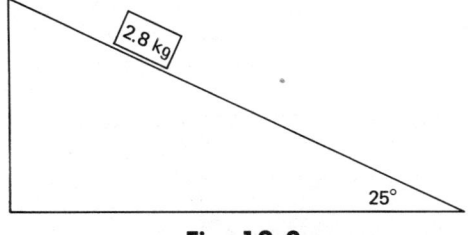

**Fig. 12-8**

## FORCE AND VELOCITY ADDITION

In adding displacements, we drew the scale model arrows in sequence from head to toe. The problem then was to find the resultant displacement. In the static force problems of introductory physics, the forces are often acting at a point. The object on which they act is in equilibrium. Therefore, the vector sum of the forces must equal 0. The scale model arrows representing the forces are shown with their tails at the equilibrium point, with the forces all pulling away from that point.

Figure 12-9 shows two equilibrium situations where three forces are acting at a point. In each case the question is, what is the magnitude and direction of the third force that just balances the other two?

We have solved the first problem; you solve the second one.

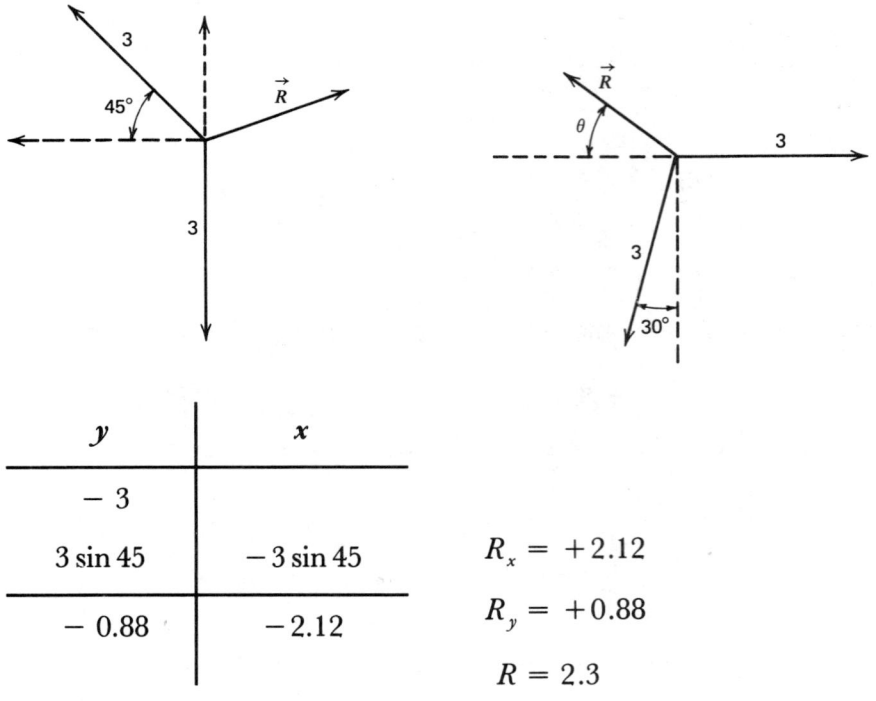

**Fig. 12-9**

If you are in a train traveling 10 mph, and walk toward the back at 4 mph, how fast are you going? The question is meaningless unless we ask, "How fast are you going with respect to a particular observer?" For instance, if somebody is walking along with you, your speed with respect to that person is 0. With respect to the train, your speed is $-4$ mph (negative, because you are walking toward the back). Your speed with respect to a person on the station platform is $+6$ mph. At least, that's the speed *if* velocities add linearly. Here's another case where you might think that it is obvious that two quantities add in a linear fashion. As a matter of fact, linearity in this case is only a very good approximation. The true addition of speeds along the same direction is given by the special theory of relativity as $w = \dfrac{u + v}{1 + uv/c^2}$. In our example, $u$ would correspond to the 10 mph of the train, $v$ would correspond to your walking speed of $-4$ mph, $w$ is the resultant speed with respect to a person on the train platform, and $c$ is the speed of light. Since the speed of light is $3 \times 10^8$ m/s, or 186,000 miles/s, the relativistic form of the equation is not needed for simple calculations of walking around on trains. (Still, this situation provides a delightful warning that in even the simplest and most obvious of relationships in introductory physics, we can not take anything for granted. The research frontier is never far away.)

Now, back on that train, suppose you walk at 4 mph from one side of the car to the other, as shown in Fig. 12-10. What is your velocity with respect to the man on the railroad platform? This situation looks like another vector problem. Indeed it is, *providing* velocities combine as vectors. Experimentally, they do, and we even have special names to distinguish between the vector, velocity, and the scalar speed. In everyday life, and in many cases even in technical work, we use these two terms interchangeably. However, technically, velocity is a vector quantity with magnitude and direction, and speed is a scalar, simply the magnitude of the velocity. In the train situation, shown in Fig. 12-10, we use the Pythagorean theorem to find the resultant velocity of the person on the train with respect to the person on the platform.

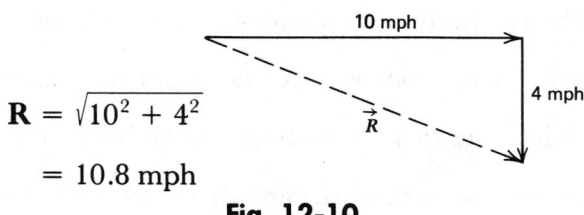

$$\mathbf{R} = \sqrt{10^2 + 4^2}$$
$$= 10.8 \text{ mph}$$

**Fig. 12-10**

Let's get off the train and onto a boat. You are in a river that flows south at 6 km/h. In still water, your small motorboat can travel with a steady speed of 10 km/h. Two different ways of getting to the other side of the river are shown in Fig. 12-11. In the first case, you resolutely keep the boat headed east, even though the current is carrying you steadily downstream. Note that two diagrams are drawn to illustrate this problem. The first sketch concentrates on the geometry of the situation, and the second is a vector diagram with arrows. It is a good idea not to get these two diagrams mixed up. In this first case, the vector diagram is simple. The resultant velocity is the hypotenuse of a right triangle with legs of 6 km/h and 10 km/h.

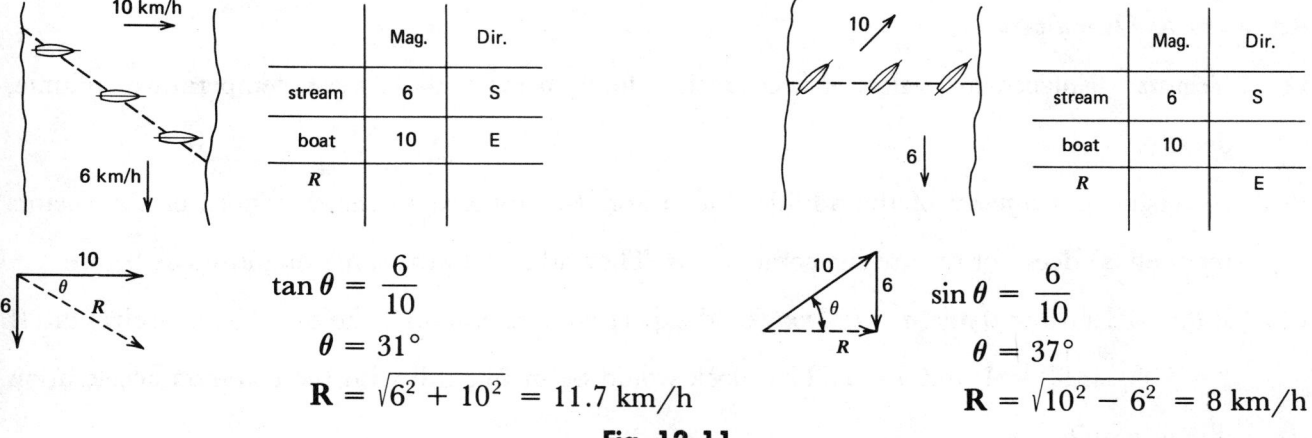

**Fig. 12-11**

The other way of getting across the river is to buck the current and head partially upstream so that your resultant velocity is due east. Of course, you won't cross the stream very fast this way, because only part of your velocity is in the eastward direction. Note that in this case the hypotenuse of the right triangle is 10 km/h and the unknown resultant velocity is one of the legs. However, the direction of the resultant velocity is known; it is due east.

In solving these stream problems, it is useful not only to sketch the geometry and to draw the vector diagrams, but to fill out a chart showing the three velocities involved and the magnitude and orientation of each. Those charts have been filled out for our two examples. Note that in each case four of the six variables are given by the problem and two are unknown. In the first case, both the magnitude and the angle of the resultant were unknown; in the second case, the angle of the boat's velocity with respect to the water was unknown. It may seem like unnecessary work to draw these diagrams and fill out the chart in order to solve such problems, but in more complicated problems, or in exam situations, the small amount of extra work can pay off by leading to clearer thinking and more direct solutions.

**Question 12-4.** Which of the two routes across the river is the faster? What are the actual times taken if the river is 5 km wide? Suppose that instead of just crossing the river, the object is to get to a point directly across from the starting point. If you use the second route, you get there directly. If you use the first route, you will end up downstream but can then turn directly upstream along the shore, bucking the current, and finally arrive at your destination. How long does it take for each of these two routes?

## Answers to Questions

12-1. Vectors: displacement, velocity, acceleration, force, area; scalars: mass, temperature, volume, density.

12-2. Reversing the *sequence* of the addition of vectors (as opposed to the directions of the vectors themselves) does not reverse the components. They add up to the same quantities as before.

12-3. If $\theta = 0°$, $\sin 0 = 0$ and $a = 0$. We would expect no acceleration if the plane were horizontal. If $\theta = 90°$, $\sin 90 = 1$ and $a = g$. The block would be in free fall with the common acceleration due to gravity.

12-4. The time taken by the second route is (5 km)/(8 km/h) = 0.63 h. Using the first route you get to the other side in only (5 km)/(10 km/h) = 0.5 h. However, you are then (6 km/h)(0.5 h) = 3 km downstream. You would have to fight your way upstream at 4 km/h. This route would take much longer. Is there yet a faster way?

## PROBLEMS

1. A displacement is 6 m at 30° north of west. What are the north and east components?

2. If you move a coin 3 cm in the $+x$ direction, 4 cm in the $-y$ direction, 2 cm in the $-x$ direction, and 3 cm in the $+y$ direction, what is the final displacement distance and direction from the origin?

3. A person moves 10 m N, 8 m 30° S of W, 6 m at 45° S of E, and 2 m E. What is the magnitude and direction of the final displacement?

4. A sled is being pulled forward by a rope, at 30° from the horizontal, which has a tension of 20 N. It is being pulled backward by a rope at an angle of 45° from the horizontal which also has a tension of 20 N. If the mass of the sled is 50 kg (and ignoring friction), what is the acceleration?

5. A block with a mass of 5 kg is on a frictionless plane that is inclined at an angle of 20° from the horizontal, as shown in Fig. 12-12. What is the acceleration of the block? (Don't forget to draw the vectors and components.)

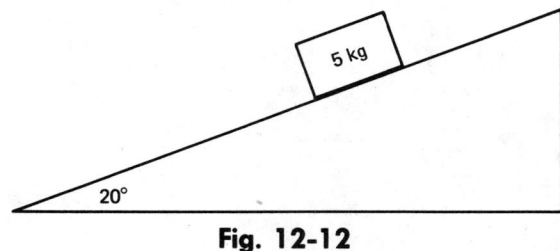

Fig. 12-12

6. Find the magnitude and direction of the force that will produce equilibrium in the situation shown in Fig. 12-13.

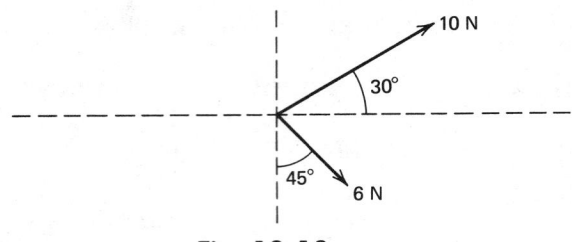

**Fig. 12-13**

7. A fighter plane flying at 1000 km/h fires a shell with a muzzle speed of 1000 km/h. How fast is the shell going with respect to the ground if it is launched in the forward direction? The backward direction? Sideways?

8. An airplane with an air speed of 200 km/h maintains a bearing due east in a wind of 100 km/h blowing from the north east (45° N of E). What is the resultant ground velocity of the plane?

# CHAPTER 13
# MOMENTUM

Can you stop a bullet with your hand? Sure—if someone tosses it to you! But not if it is traveling with a normal muzzle velocity of 500 m/s. There is something about the *speed* of an object that characterizes its motion.

Can you stop a car from rolling if it is traveling at 5 mph? Sure—if it's a toy car, or perhaps even if it's a compact. But not if it's a bull Cadillac. There is something about the *mass* of a moving object that characterizes its motion.

The product of *mass* and *velocity* is called *momentum*.

$$\text{momentum} = mv$$

Evidently, momentum has something to do with how hard it is to start or stop something. Note that we might have named other combinations of $m$ and $v$: $m^2v^3$, for instance. It turns out that the simplest combination $mv$ is very useful and has a vital conservation property.

**Question 13-1.** What is the momentum of a ball which has a mass of 100 g and is traveling at 30 m/s?

In spite of the importance of momentum, no proper name has been assigned to its unit. The unit is simply kg(m/s).

## THE RELATIONSHIP BETWEEN FORCE AND MOMENTUM

Newton's second law of motion relates the force on an object to the acceleration produced: $F = ma$. Note that $a = \Delta v/\Delta t$. Newton's second law could be written $F = m(\Delta v/\Delta t)$. If the mass of the object is constant, we can move the mass inside the parentheses: $F = \Delta(mv)/\Delta t$. It appears that force is related to the change in momentum, not just the change in velocity. *Indeed, this is the most general form of*

*Newton's second law:*

$$F = \Delta(\text{momentum})/\Delta t$$

The net force acting on an object is equal to the *time rate of change of the momentum of the object.*

**Question 13-2.** What happens if the mass of the object changes?

To produce a particular change of momentum, a small force could be applied to an object for a long time, or a large force for a small time.

$$\Delta(mv) = F\Delta t$$

The product of force and time during which it is applied is called *impulse.*

Let's take the case of a bullet with a mass of 5 g and a speed of 500 m/s. Its momentum is $mv = (5 \times 10^{-3} \text{ kg})(500 \text{ m/s}) = 2.5 \text{ kg(m/s)}$. If it lodges in a bone and stops, it must come to rest in a distance of a few centimeters. Its average velocity while stopping would be 250 m/s. Therefore the time of stopping would be $\Delta t = (2.5 \text{ cm})/(250 \text{ m/s}) = (2.5 \times 10^{-2})/(2.5 \times 10^{2}) = 1 \times 10^{-4}$ s. The stopping *impulse* must be equal to the change of momentum: 2.5 kg(m/s) = 2.5 Ns. The force during this short interval is $F = (2.5 \text{ Ns})/(1 \times 10^{-4} \text{ s}) = 2.5 \times 10^{4}$ N. This quick, sharp blow will break the bone.

Whenever a force is applied to an object for a given time, the momentum of the object changes. If the force is constant during the time, as shown in Fig. 13-1, the calculation of the change in momentum is simple: $\Delta(mv) = F\Delta t$. If the force is changing with time, as in Fig. 13-2, we can plot $F(t)$, force as a function of time. *The area under the $F(t)$ curve is equal to $\Delta(mv)$.*

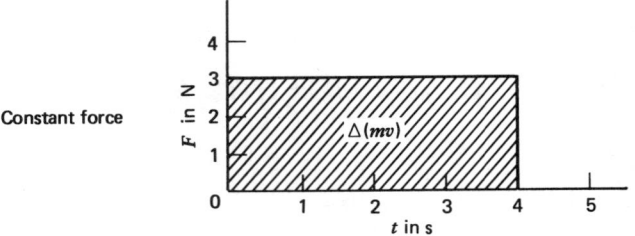

**Fig. 13-1** The change of momentum is equal to (3 N) × (4 s) = 12 Ns = 12 kg(m/s).

**Question 13-3.** In this example the mass of the object was not given. How can you find the resultant velocity change?

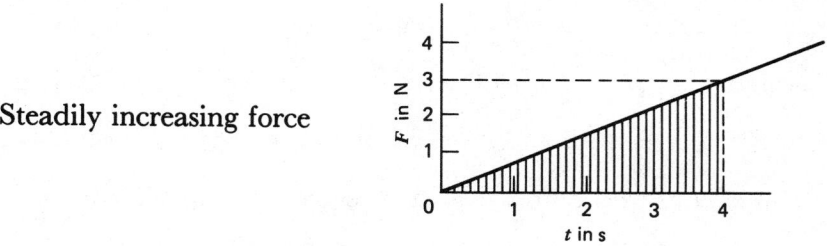

Steadily increasing force

**Fig. 13-2** The average force during the first 4 s is 1.5 N. Therefore $\Delta(mv) = (1.5 \text{ N})(4 \text{ s}) = 6$ Ns. This value is also the area under the $F(t)$ curve.

$$\left[\text{area of a triangle equals } \tfrac{1}{2}(\text{base}) \times (\text{height})\right]$$

The $F(t)$ curves in Fig. 13-3, drawn on the same graph, explain the purpose and usefulness of seat belts. The problem for the occupants in a car collision is what happens during the second collision when *they* are brought to a stop. Their momentum will decrease to zero, one way or another. The time interval involved is a matter of life or death.

Assume that you drive at 40 mph into a concrete wall. The car will stop in a distance of less than 1 m, that distance being taken up by the space that the motor and things used to occupy. Assuming that the car slows down with constant acceleration from an initial speed of 40 mph, which is about 18 m/s, the time it takes is equal to

$$\Delta t = \Delta x/v_{av} = (1 \text{ m})/(9 \text{ ms}) = 0.11 \text{ s}$$

If your seat belts are well adjusted and hold, you will slow down in that same time. If you are not

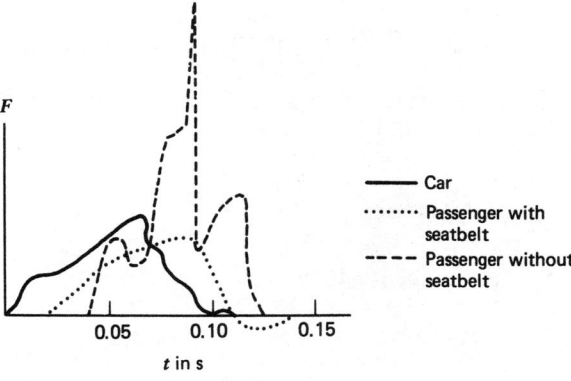

**Fig. 13-3**

wearing seat belts, you will feel no strain until your body hits the steering wheel. The wheel will not take long to collapse, giving you more free time before you hit the dashboard and windshield. Then you will stop in a hurry unless you make it all the way through the windshield. On the graph of force versus time for this accident, we have drawn three curves: one for the car frame, one for a driver with seat belts, and one for the driver without seat belts. The time scale of events is realistic in terms of automobile manufacturers' tests with dummies. Actual events vary drastically, of course, depending on whether your stomach gets caught in the wheel and on which portion of your head hits the glass.

There is one important thing to keep in mind during the tenth of a second that all this action takes place. The change of momentum of your body will be the same regardless of whether you are wearing a seat belt. You were traveling at 18 m/s, and about one-tenth of a second later your speed is zero. Therefore, the total area under the $F(t)$ curve must be the same

$$\Delta(mv) = (60 \text{ kg})(0 - 18 \text{ m/s}) = -1.1 \times 10^3 \text{ kg(m/s)} = -1.1 \times 10^3 \text{ Ns}$$

If you could stretch the impulse out evenly over the time of the collision, the force would be $F = (-1.1 \times 10^3 \text{ Ns})/(0.11 \text{ s}) = -1 \times 10^4 \text{ N}$. The minus sign just indicates that the direction of the force is opposite to your original direction.

As you can see from the graph in Fig. 13-3, the actual effect of the seat belt is not so ideally provident, but the maximum $g$ produced is tolerable. Besides, the seat belt exerts its restraining forces on sections of the body that are more flexible than the skull, spreading the force out over a larger area of the body.

Note that it is not exactly the car collision that kills the driver. Instead, the fatal blow comes from the second collision when the driver strikes the inside of the car. The car has pretty much come to rest before the unbelted driver hits. The driver's stopping distance, and thus the stopping time, is shorter than that of the car by a factor of about 10. Consequently, the stopping force on the body is about 10 times what it would have been had the driver been fastened to the car.

## THE CONSERVED NATURE OF MOMENTUM

If no external forces act on an object, the *change* of momentum must be zero: $\Delta(mv) = 0$ if $F_{\text{ext}} = 0$.

This conservation property of momentum is what makes the momentum of a system such an important quantity. Let's see some of the surprising facts that occur because of conservation of momentum.

1. A ball flies along at constant horizontal speed, but eventually hits the ground and stops.

   **Question 13-4.** Apparently momentum wasn't conserved. What's wrong with the law?

2. A hockey puck slides along the ice at almost constant velocity.

   **Question 13-5.** Since gravity is acting on the puck, how can we apply the law of conservation of momentum? The puck is not isolated from external forces.

3. The sliding puck hits an identical one head-on. The target puck was standing still originally. After the collision it takes off with the original velocity, leaving the original one motionless. This is an astonishing sight! First note that momentum of the whole system is conserved. The original momentum was $m_1 v + m_2(0)$. (The second puck had zero speed.) After the collision, the momentum of the whole system was $m_1(0) + m_2 v$. Since $m_1 = m_2$, the momentum after equals the momentum before:

$$m_1 v + m_2(0) = m_1(0) + m_2 v$$

4. Suppose that two identical pucks are tied together to begin with, but with a compressed spring between them. The original momentum is zero because the whole system is just sitting there on the ice. If the tie is broken, the pucks will spring apart. They will have equal speeds but exactly opposite directions.

   **Question 13-6.** Now apparently the momentum conservation law fails. The original momentum of the system was zero. Afterwards each puck has momentum $mv$. What is conserved?

5. Here's a case, shown in Fig. 13-4, where a particle (such as a proton) might hit an identical particle (another proton) in a glancing collision. The target proton with mass $m_2$ was standing still. The original momentum was in the $x$ direction and must be conserved. Its value was $m_1 v_0$. The momentum of $m_1$ afterwards was $m_1 v_1$. The $x$ component of that momentum is $m_1 v_1 \cos \theta$. The $x$ component of the momentum of the target proton after the collision is $m_2 v_2 \cos \phi$. To conserve the $x$ component of momentum

$$m_1 v_0 = m_1 v_1 \cos \theta + m_2 v_2 \cos \phi$$

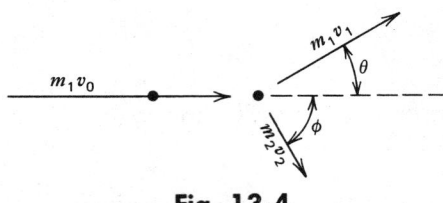

Fig. 13-4

The original $y$ component of momentum was zero (by definition of the coordinate system). Therefore to conserve the $y$ component of momentum

$$0 = m_1 v_1 \sin \theta - m_2 v_2 \sin \phi$$

The $+y$ component of momentum of proton 1 must be equal in magnitude to the $-y$ component of proton 2.

6. Instead of thinking about two pucks on ice, consider two skaters facing and shoving each other apart. One of them has twice the mass of the other—say 50 kg and 100 kg. What happens if friction is negligible so that momentum is conserved?

$$\text{original momentum} = \text{final momentum}$$
$$0 = m_1 v_1 + m_2 v_2$$
$$0 = (50 \text{ kg})v_1 + (100 \text{ kg})v_2$$
$$v_1 = -2v_2$$

The light skater glides backward with twice the speed of the heavy one.

**Question 13-7.** Consider the "system" of a tennis ball and the earth. Drop the tennis ball and its speed (and momentum) increases. Then it hits and its velocity goes to zero and reverses. How can momentum be conserved?

## Answers to Questions

13-1. We should use our standard SI units: momentum $= (0.1 \text{ kg})(30 \text{ m/s}) = 3 \text{ kg(m/s)}$.

13-2. Yes, the mass of an object can change. The mass of an object actually depends on its speed, although the effect is very small unless the speed is close to that of light. There are less exotic situations where the mass of an object can change. For instance, a rocket shoots a major share of its mass out the back end. In all these cases, the exact equation for Newton's second law is in terms of the time rate of change of momentum. $F = ma$ is an approximation!

13-3. A constant force of 3 N applied for 4 s to any object produces the same 12 Ns of additional momentum. The change of *velocity* produced does depend on the mass of the object. If the object has a mass of 1 kg, the change of velocity is 12 m/s. If the object has a mass of 1 g, the change of velocity is $12 \times 10^3$ m/s.

13-4. Nothing. The law requires that the object must be isolated, with no external forces acting on it. In this case, gravity was pulling the ball downward, increasing its vertical momentum. When the ball hit the ground, other forces acted on it.

13-5. The weight of the puck is supported by the ice. The vertical forces on the puck cancel, leaving only a very small horizontal friction force. The puck, to first approximation, is not subject to any *net* external force.

13-6. *Momentum is a vector*. If the velocities are equal but opposite, and $m_1 = m_2$, then the final momentum of the system remains zero

$$m_1(0) + m_2(0) = m_1 v + m_2(-v) \quad \text{and} \quad m_1 = m_2$$

13-7. While the ball is dropping to the earth, the earth is rising to meet the ball. Consider, however, the velocity of the earth!

$$m_1 v_1 + M_2 V_2 = 0$$

$$M_2/m_1 = (6 \times 10^{24} \text{ kg})/(0.030 \text{ kg})$$

After the collision, the earth and ball spring apart from each other—not with the speeds they had just before the collision, however. Energy has been lost, a subject we study in the next chapter. However, since the original momentum of the earth–ball system was zero, it remains zero throughout the action.

## PROBLEMS

1. An ice skater (or roller skater) stands still facing another skater gliding toward her with a speed of 2 m/s. Her mass is 50 kg; his is 70 kg. If they embrace (and do not stop their motion by digging their skates into the ice), what is their resultant joint velocity?

2. On an air hockey table, a floating disk with a speed of 4 m/s strikes another disk with equal mass. The first disk glances off at an angle of 30° from its original direction with a speed of $v_1$; the second disk is knocked at an angle of 60° from that original direction with a speed of $v_2$. What are the values of $v_1$ and $v_2$?

3. In a head-on collision, a proton (relative mass 1) traveling at $3 \times 10^7$ m/s strikes the nucleus of a helium atom (relative mass 4) at rest and bounces directly backward with a speed of $1.8 \times 10^7$ m/s. What is the velocity of the helium nucleus?

4. In the radioactive decay of uranium, an alpha particle (helium nucleus), with a mass of about 4 on the atomic scale, is emitted with a velocity of $1.5 \times 10^7$ m/s. What is the recoil velocity of the remaining atom of thorium, which has a mass of about 234 on the same scale?

5. A child with a mass of 22 kg running at 2.5 m/s jumps onto a 12-kg wagon from the rear. What is the resultant velocity of wagon and child?

6. A 1.8-kg ball hits a wall straight on at 6.5 m/s and bounces off at 4.8 m/s. What impulse is experienced by the ball?

7. A 12-kg wagon is pushed with a force of 7.0 N for 1.5 s, then with 4.5 N for 1.2 s, then with 10.0 N for 2.0 s. (a) What is the total impulse applied to the wagon? (b) What is the wagon's change of velocity (assuming that rolling friction is negligible)?

8. Suppose that the graphs in Fig. 13-5 show the force as a function of time exerted by a karate expert and by a boxer. (a) What is the impulse delivered in each case? (b) Speculate about which would be the more dangerous blow. Which would send you flying across a room? Which would be more apt to break a bone?

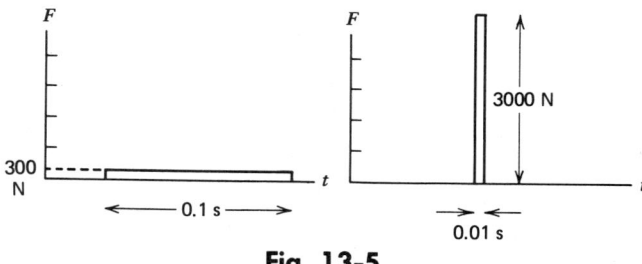

**Fig. 13-5**

9. A 5.0 kg rifle, suspended on strings, fires a 4.0-g slug with a muzzle velocity of 520 m/s. What is the recoil velocity of the rifle?

10. For the passenger in a car accident, the *change* of velocity is a crucial factor. Suppose that on an icy road a station wagon with a mass of 2000 kg, traveling at 30 mph, runs head-on into a 1000 kg compact that was traveling at 30 mph in the opposite direction. The cars lock together. Assume that on the icy road they slide without friction. What is the change of velocity of each car?

# CHAPTER 14
# ENERGY AND THE MULTIPLICATION OF VECTORS

When you multiply 3 by 4 you are really just adding three 4's together. What possible meaning could there be to multiplying two *vectors* together? If you multiply 3 m times 4 N, you cannot very well add 4 N with the aid of 3 m. Nevertheless, we *can* multiply vectors, and the process is one more example of the way mathematics can be used as a model for real-world phenomena. Vector multiplication starts out as a defined mathematical operation. Then you look around and see if the results describe some process or phenomenon.

As a matter of fact, consistent mathematical procedures have been worked out for two kinds of vector multiplication. Each is applicable to a particular group of physical processes. First, we'll take up the *dot*, or *scalar*, product, and then we'll study the *cross*, or *vector*, product. In each case, however, we will consider a particular physical phenomenon that requires mathematical description.

The first case will require *work*. By "work" we don't just mean hard studying. We mean the technical expression *work*. The technical expression ought to agree with our everyday meaning of the word, if at all possible. When you do work, you expect to get paid. If a machine does work, it must be supplied with fuel.

**Question 14-1.** Let's take a specific example. How much would you expect to get paid for holding a 10-lb box at waist height for 2 h? Note that in order to do this you would have to exert a continuous upward force of about 50 N.

There is a difference between the energy that you expend in pushing or holding something and the work that is done on the object. When you shove or hold something your muscles may be twitching back and forth even though the object may remain motionless. Because your muscles are moving, energy is being used up, you are getting tired, and you ought to be paid. As far as the process is concerned, however, there may be an alternative way of producing the same effect without doing any work at all. For instance, if you are gluing two boards together and want to clamp them tightly for half an hour, you could tire yourself out by pressing the boards with your hands for that time. Alternatively, you could clamp the boards together with a C-clamp, which would exert a large force and yet require no fuel and no money. No work has been done on the boards.

The crucial variables involved in work are *force* and *distance through which the object moves because the force is applied*. The greater the force or the greater the distance, the more work is done. In our formal definition, *work is the product of force and distance*. If 1 N is exerted through a distance of 1 m, then unit work has been done. The name of the unit is the joule (J):

$$1 \text{ J} = (1 \text{ N})(1 \text{ m})$$

**Question 14-2.** Here we have a product of two vectors. So far, however, it looks like an ordinary product of two numbers. What's different, just because the variables are vectors? For instance, suppose you exert a force of 50 N to support a box and carry it at waist height for a distance of 10 m. How much work have you done?

In our word definition of work, there was a small but crucial proviso. The distance involved is the distance *through which the object moves because the force is being applied*. Another way of defining the same thing is to say that the work done is equal to the product of the distance through which something moves and the *effective* force in that direction. Figure 14-1 shows a force being exerted down on a box at an angle of 30° from the horizontal. $F \sin \theta$, the vertical component, is just pushing the box into the floor. $F \cos \theta$, the horizontal component, is the effective force in moving the box horizontally. If the box is shoved $x$ m, then the work done is $W = (F \cos \theta)(x)$.

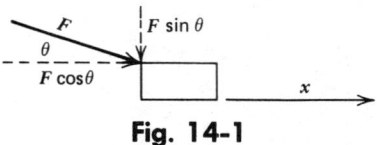

**Fig. 14-1**

The dot product, or scalar product, of two vectors is defined to be the product of the magnitudes of the two vectors and the cosine of the angle between them. This is just the definition we need to describe the role of force and displacement in the production of work. If **A** and **B** are vectors, then the dot product, or scalar product, is defined to be

$$C = \mathbf{A} \cdot \mathbf{B} = |A||B| \cos \theta$$

The product quantity $C$ is a scalar, hence, the name, scalar product. The symbol for this particular kind of multiplication is the dot between the two vectors, hence, the name, dot product. In these terms, the definition of work is

$$W = \mathbf{F} \cdot \mathbf{x} = |F||x| \cos \theta$$

Work is indeed represented by a scalar quantity. There is no north, east, south, or west direction to it.

**Question 14-3.** The force required to extend a seat belt spring is almost independent of the extension. Suppose you exert a constant force of 10 N to pull out the seat belt 50 cm. Your force is in the direction of the extension. How much work did you do? Now let the spring pull the seat belt back slowly through 50 cm. The force that it is exerting must be 10 N, and since you are resisting the motion, you must be still pulling with approximately 10 N in the outward direction (otherwise, the seat belt would go snapping back). On the return trip, you are still pulling outward with a force of 10 N, but now the seat belt is retracted in the opposite direction through a distance of 50 cm. Now how much work do you do?

As we have seen, with ordinary springs the restoring force is proportional to the stretch length $F_{\text{restoring}} = -kx$. The minus sign reminds us that the restoring force is in the opposite direction from the extension. Suppose that you have a spring with a constant $k = 100$ N/m. If you stretch the spring 10 cm, how much work have you done? To be sure, $W = \mathbf{F} \cdot \mathbf{x}$, and in this case, $\mathbf{F}$ and $\mathbf{x}$ are in the same direction. However, $\mathbf{F}$ is not constant. On the following graph, plot $F$ as a function of $x$ for $x = 0$ to $x = 10$ cm. Label the axes so that the graph agrees with the value of the spring constant that we have given. [Note that the spring constant is the slope of the $F(x)$ graph.] We have already seen how to take the product of two variables when one is graphed as a function of the other. For instance, when we plotted $v(t)$ for motion, the area under the curve for a given interval was equal to the distance traveled during that interval. In the same way, the area under the graph of $F(x)$ is the work done for an extension $x$. In this particular case, the area under the curve is a triangle. How much work does it represent? _____.

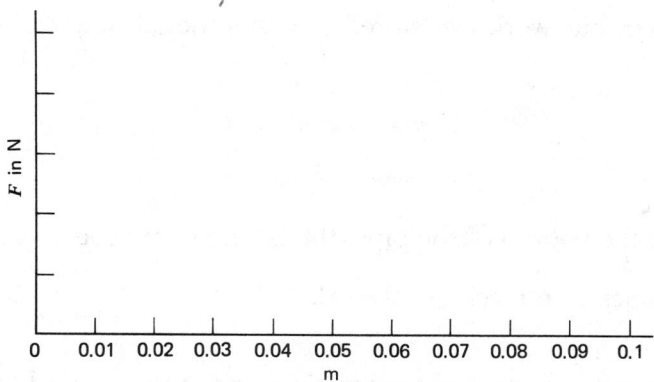

**Question 14-4.** What has happened to our dimensions in this situation? Should the area of the graph have dimensions $L^2$? What are the dimensions of work? Is there a contradiction?

## WORK AND ENERGY

So far we have defined work in terms of the dot product of force and displacement. At the same time we talked about expending (using up) energy in order to do the work. Energy is another of the conserved quantities in the world. If you have an isolated system its energy will remain constant. However, energy can take many forms. In some cases it seems as if the visible effects of energy disappear, leaving nothing. In each case, however, we can account for the missing energy.

Suppose you do work by exerting a force to lift a large quantity of water to the top of a hill. The work done is $\mathbf{F} \cdot \mathbf{y}$, where $\mathbf{y}$ is the displacement. The minimum amount of work required is $mgh$, where $mg$ is the weight of the water and $h$ is the vertical height.

**Question 14-5.** Why is this the *minimum* work required? Can $\mathbf{F} \cdot \mathbf{y}$ be greater? If so, what happens to the extra work?

After the water is at the top of the hill, it is not obvious that it has energy. Our work seems to have disappeared. However if we let the water drop down a pipe, it falls faster and faster. We define motion energy as a function of mass and velocity—but a different combination than that of momentum!

$$\text{motion, or kinetic energy} = \tfrac{1}{2}mv^2$$

In the case of the water, our work was stored in gravitational potential energy, and then turned into kinetic energy

$$\mathbf{F} \cdot \mathbf{y} = mgh + \text{heat}$$

$$mgh \rightarrow \tfrac{1}{2}mv^2$$

If there is a turbine at the bottom of the pipe, the falling water could turn a mill, thus converting part of the kinetic energy back into mechanical work.

**Question 14-6.** Why can only part of the kinetic energy be converted back into mechanical work?

The turbine might turn a generator to produce electric energy, which in turn, could produce many other forms of energy. So far in our scientific studies of phenomena, we have always been able to account for missing energy in any transformation. The conservation of energy is one of the great cornerstone laws of our science.

## THE SIZE OF THE JOULE

The SI unit of work or energy is the *joule*.

$$1 \text{ N·m} = 1 \text{ J}$$

To get an idea of the size of the joule, raise this book.

**Question 14-7.** How high? Well, how much does the book weigh? Heft it and estimate the mass in kg or weight in N. (On page 27 you were asked to find the mass of this text.) The work to lift the book is $\mathbf{F} \cdot \mathbf{y} = mgh$. How high should you raise the book in order to do one joule of work?

Here are approximate values for some familiar expenditures of energy:

a. A standard two-cell flashlight uses about 1 J in 1 s.

b. A 100-watt (W) bulb uses 100 J in 1 s.

c. It takes 4 J to raise the temperature of 1 g of water by 1 C°.

d. It takes 64,000 J to raise the temperature of a cup of water from room temperature to boiling.

e. In friendly ping pong games, the ball rarely has more than 0.1 or 0.2 J of kinetic energy.

f. A diet of 2000 food calories (2000 kilocalories) requires ingesting about $8 \times 10^6$ J.

Here's how to calculate some of these relationships. The calorie is also a unit of energy. The food calorie ≈ 4000 J. The unit of *power* is the watt (W). Power is the time rate of using energy.

$$1 \text{ W} = 1 \text{ J/s}$$

Consequently, a 60-W bulb uses 60 J every second.

**Question 14-8.** How much energy would you expend in running up one flight of stairs at home? (Standard distance between floors in a house ≈ 3 m.) How long would it take you? (Measure or estimate.) How much power do you use?

## OTHER DOT PRODUCTS

There are many other dot products useful in physics. Power, for instance, which is the time rate of doing work, equals **F·v**. When you study electricity and magnetism you will have to describe dipoles. An ordinary compass needle, for instance, is a dipole with a north pole on one end and a south pole on the other. It takes work, or energy, to twist a magnetic dipole out of line with a magnetic field or an electric dipole out of line with an electric field. The stored energy of an electric dipole is defined as $U = -\mathbf{E} \cdot \mathbf{p}$. As shown in Fig. 14-2, the dipole strength and direction is represented as a vector **p**, the direction from the minus pole to the positive pole. The electric field is represented by the vector **E**.

**Question 14-9.** According to this expression for the stored electric energy of a dipole, the energy is negative when the dipole is lined up with the field. How can there be negative energy? Under what circumstances would the energy be maximum positive? Under what circumstances would the energy be zero?

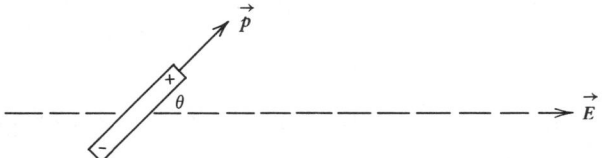

**Fig. 14-2**

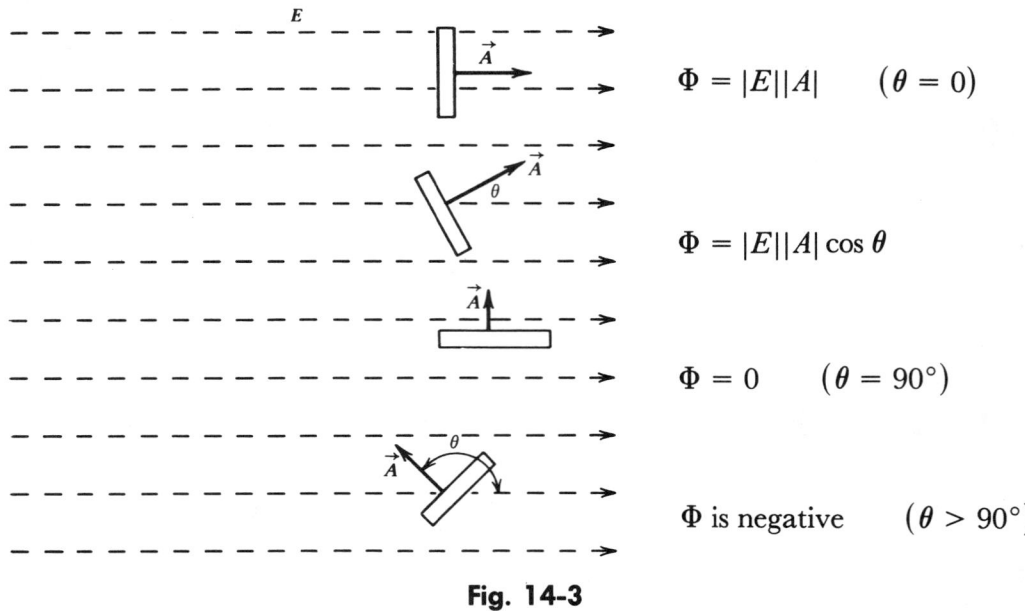

**Fig. 14-3**

Another use of the dot product is illustrated in Fig. 14-3. If some influence, like electric field or radiation (perhaps light), is streaming over a region, the total "flux" or "illumination" depends on the intensity of the influence, the area covered, and the orientation of the surface area to the direction of the influence. In the case of an electric field, the electric flux is given by

$$\Phi = \mathbf{E} \cdot \mathbf{A}$$

**Question 14-10.** But since when can area be represented by a vector? What are the three conditions that such a representation must satisfy?

Note how convenient it is to represent area by a vector. The maximum electric flux occurs when the electric field is in line with the normal (perpendicular) to the surface. If the normal to the surface were perpendicular to the electric field, then no electric flux would penetrate the surface. The electric field lines would just be skimming along the surface. If the normal to the surface were pointed back along the direction of the electric field, then the flux would be negative. The positive or negative feature simply indicates whether the electric field is going in to or out of a surface. The outward direction of the surface is usually arbitrary, except that on a curved surface, like that of a sphere, the outward direction is rather obvious.

**Handling the Phenomena** You can observe the influence of the dot product on illumination by looking at a white piece of stiff paper or cardboard that is illuminated by a single lightbulb. At some distance from the bulb, hold the sheet so that its normal points back toward the bulb. Note how bright the paper looks. Now turn the paper slowly until finally the normal from the surface is perpendicular to the direction back to the bulb. Keep looking at the paper as you turn it, and note how much darker it becomes. As you turn it, of course, it is intercepting progressively smaller amounts of the light.

## THE CROSS, OR VECTOR, PRODUCT

There is another mathematically consistent way to define the multiplication of vectors. It turns out that this method is also very useful for describing many physical phenomena. The symbol for this type of multiplication is a multiplication sign: $\mathbf{C} = \mathbf{A} \times \mathbf{B}$. Note that we indicated that the product is also a vector. The process is therefore crucially different from the dot, or scalar, product.

**Handling the Phenomena**  To see why we need a vector product, just open a door. The heavier the door, the more obvious the point. To open a heavy door, you must exert a force. You exert the force through some distance and so do work. But you also have to make a choice about how and where you apply the force. Try opening a door by pushing on the hinges. Try opening a door by pulling on a handle, but along the same direction as the door (radially from the hinges). Draw an arrow on the diagram in Fig. 14-4 showing the best way to exert a force to open a heavy door. (Don't just use your imagination. Try shoving and pulling on a door at odd angles and positions. Of course, if in order to be polite, you have ever tried to open a heavy door for someone who is standing in front of you, you know the problem.)

If you want to push something in a straight line, you exert a force; if you want to twist something or turn it in a circle, you must exert a *torque*. The effectiveness of your effort to turn something depends

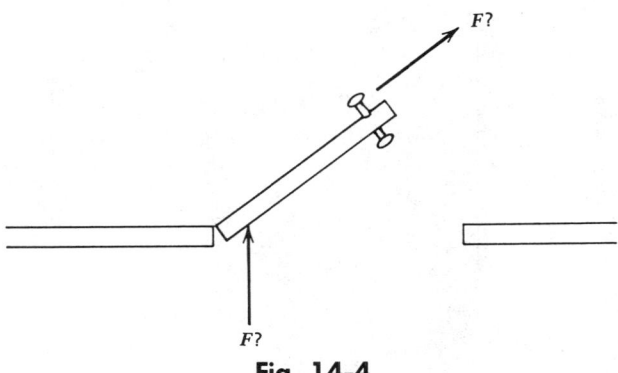

**Fig. 14-4**

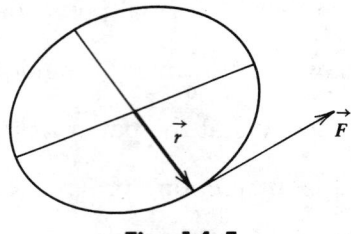

**Fig. 14-5**

on the force you exert *and* the *lever arm* you use. In Fig. 14-5, the lever arm is labeled **r** because it is radial. The torque is defined to be

$$\tau(\text{tau}) = \mathbf{r} \times \mathbf{F}$$

To maximize torque, you should maximize the length of the lever arm |**r**|, the magnitude of the force |**F**|, *and* the direction of the force should be perpendicular to the direction of the lever arm. The magnitude of the cross product is

$$|\tau| = |\mathbf{r}||\mathbf{F}| \sin \theta$$

**Question 14-11.** Does this last formula agree with your experience in opening doors? What happens if $\theta = 0°$? If $\theta = 45°$? If $\theta = 180°$? If $\theta = 270°$?

Note that with a dot product of force and displacement, the effective force in producing work is $F \cos \theta$, the component of **F** in the direction of the displacement. In the case of torque, the effective force is $F \sin \theta$, the component of **F** *perpendicular* to the lever arm.

Since torque is a vector, it must have a direction as well as a magnitude. Figure 14-6 shows the assigned direction. It is axial, along a direction perpendicular to the other two vectors. The torque you

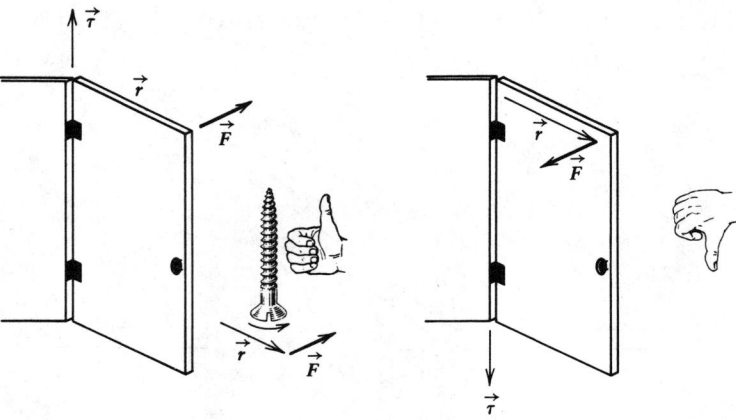

**Fig. 14-6**

exert in opening a door is therefore vertical, directed along the hinges. You can determine whether the direction is up or down by curling the fingers of your right hand in the direction of the motion of the door. Your thumb then points in the direction of the torque. Another way to determine the direction of a cross product is shown in the diagram. Imagine that you are turning a right-hand screw from the first vector to the second; for instance, from **r** through **F**. The direction of the advancing screw is the direction of the vector product.

Since the direction of the vector product depends on the arrangement of the other two vectors, we have **A** × **B** ≠ **B** × **A**. Instead, we have (**A** × **B**) = −(**B** × **A**). The reason for this is shown in Fig. 14-7. The technical statement of this property is that vector multiplication is noncommutative.

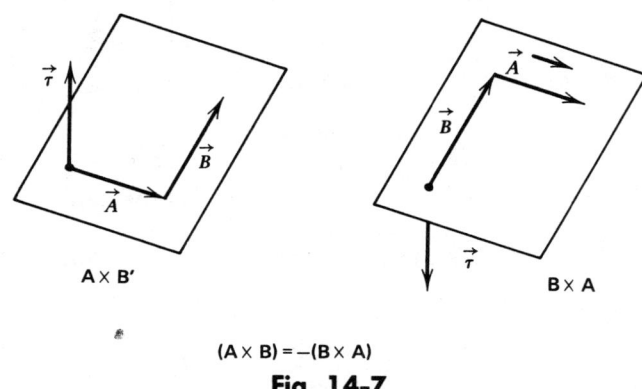

(A × B) = −(B × A)

**Fig. 14-7**

**Question 14-12.** The dot product of a force and a distance is defined to be work. A newton·meter (N·m) is a joule (J). The cross product of a force and a length is defined to be torque. Torque seems to have the same units as work. Does that mean that torque is the same thing as work?

There are many other uses for the vector product in physics. If a particle with electric charge $q$ is traveling with velocity **v** in a magnetic field **B**, then the force on the particle is

$$\mathbf{F} = q\mathbf{v} \times \mathbf{B}$$

In Fig. 14-8, we show an example of this electromagnetic force. A magnet with a north pole on top and a south pole on the bottom produces a vertical magnetic field **B**. If a positively charged proton is shot in a horizontal vacuum chamber between the pole faces, the force exerted on the proton will be centripetal (radial toward the center). Beside the diagram of the magnet draw for yourself the arrows representing **v** and **B** and the resultant force. The centripetal force will tend to make the proton travel

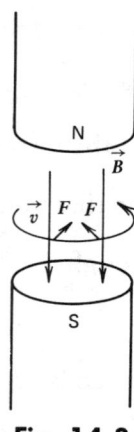

**Fig. 14-8**

in a circular path. As you look down on the path from the direction of the north pole of the magnet, will the proton be traveling clockwise or counterclockwise? ▭. If the particle were an electron instead of a proton, its charge would be negative and its path would be in the opposite rotational sense.

Do you remember we claimed that it was convenient to represent a surface area by a vector? Indeed, we can even represent the area as the vector product of two other vectors, the length and the width. In Fig. 14-9, we show the length and width of a parallelogram as vectors. The magnitude of the area is correctly given by

$$|\mathbf{A}| = |\mathbf{L}||\mathbf{W}| \sin \theta$$

**Question 14-13.** What difference would it make in defining the area of a surface if we used $\mathbf{W} \times \mathbf{L}$ instead of $\mathbf{L} \times \mathbf{W}$?

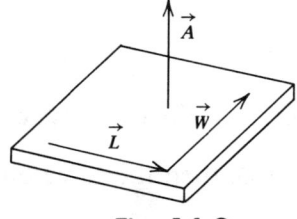

**Fig. 14-9**

There is one small technicality about vector products that we should mention, though it will probably make no difference to you in a first-year physics course. The product of vector multiplication is not really a vector; it is a pseudovector. With an ordinary vector, if you reverse the sign of all the components (e.g., $x$ goes to $-x$, $y$ goes to $-y$, $z$ goes to $-z$), the vector itself will be reversed in direction. However, if you reverse all the components of a radius vector, turning them into $-\mathbf{r}$, and all the components of a force, turning them into $-\mathbf{F}$, then the two negatives will cancel each other out and the torque will remain in its original direction. Try this for yourself. Draw $\mathbf{r} \times \mathbf{F}$ and determine the direction of the resulting torque (in or out of the page). Then draw the arrows for $(-\mathbf{r}) \times (-\mathbf{F})$ and determine the direction of the resulting torque.

## Answers to Questions

14-1. Why should I pay you for holding a 10-lb box for 2 h? I could set it on a table and keep it there all day for nothing. You may have done work, but no useful work was done on the box.

14-2. The crucial point is that work is equal to the product of force and *distance through which the force is applied*. If you exert 50 N *up*, and carry the box *horizontally*, you do no work on the box.

14-3. As you pull the seat belt out, you are doing work equal to $\mathbf{F} \cdot \mathbf{x} = (10 \text{ N})(\frac{1}{2} \text{ m}) = 5$ J. On the return trip, you are pulling outward, but the belt is moving inward. The angle between $F$ and $x$ is 180°. $\cos 180° = -1$. Therefore you do $-5$ J work. The spring does $+5$ J work.

14-4. Work has dimensions of $\left(M\dfrac{L}{T^2}\right)L = ML^2T^{-2}$. The graph area represents the product of the variables on the axes: force times length.

14-5. Instead of lifting the water straight up, we might truck it up a sloping road. We would have to exert a force large enough to lift the water *and* to overcome friction. The work done against friction makes gears and wheels hotter. The extra work can be accounted for in terms of the heat produced.

14-6. In each conversion stage, some of the useful energy is lost as heat because of friction.

14-7. The mass of the book is *about* 500 g. Therefore, the weight is about 5 N. It would take 1 J to raise the book $\frac{1}{5}$ m = 20 cm.

14-8. Multiply your weight by the vertical height. For instance,

$$mgh = (70 \text{ kg})(9.8 \text{ N/kg})(3 \text{ m}) \approx 2100 \text{ J}$$

If you can run upstairs in 3 s, your power (for that short burst) is $P = $ (energy)/(time) $=$ (2100 J)/(3 s) = 700 W. That's almost one horsepower (H.P.)! (746 W = 1 H.P.)

14-9. When a dipole is lined up with the field ($\theta = 0$), the dipole is in a trapped position; it would take positive energy to twist it. When $\theta = 90°$, $U = 0$. When $\theta = 180°$, $U = +|E||p|$.

14-10. To be a vector, a quantity must have magnitude, direction, and combine like a displacement. An area has magnitude ($m^2$), and can be assigned the direction of its normal (perpendicular), as shown in the diagram. As you can see in this example, area can be represented by a vector.

14-11. If $\theta = 0°$, you are pulling on the handle along the line of the door. $\sin 0° = 0$, and your torque is zero. You cannot open the door. If $\theta = 45°$, your *effective* force is $F \sin 45 = 0.7\ F$. If $\theta = 180°$, you are pushing on the handle along the line of the door, back toward the hinges; torque is zero. If $\theta = 270°$, your torque is in the opposite direction from what it is when $\theta = 90°$. You are closing the door.

14-12. The quantities of torque and work are radically different. Torque is a vector; work is a scalar.

14-13. Since $(\mathbf{W} \times \mathbf{L}) = -(\mathbf{L} \times \mathbf{W})$, the order of multiplication defines which surface is top and which is bottom.

# CHAPTER 15
# THE SINUSOIDAL FUNCTION: $y = A \sin \theta$

The power functions describe certain types of phenomena where, as one variable increases, the other variable increases or decreases without limit. However, there are many phenomena in this world that are repetitive, or cyclic. As time goes on, a pendulum bob swings back and forth, but never keeps going far in any one direction. The tides rise and fall with a regular rhythm, but the water never goes away completely. The voltage in your electric wall outlet increases and decreases 60 times a second, but never goes more positive or negative than a preset maximum. In order to describe phenomena like these, we need a mathematical function that also oscillates, or rises and falls.

You have already made a pendulum and found the functional dependence of the period on the length of the string. Consider now the *position* of the pendulum bob as a function of *time*. In Fig. 15-1, we have labeled the central position as zero, with positions to the right being positive and to the left being negative. If we plot the distance from the zero point as a function of time, we can enter a number of points without even making precise measurements. Let's imagine that we start our stopwatch when the pendulum bob is sweeping through zero. Then, at some later time, the bob is at a maximum distance to the right, which we record with the second point on the graph. The bob then reverses direction and in the same time is going back through zero. We record that event with the third point on

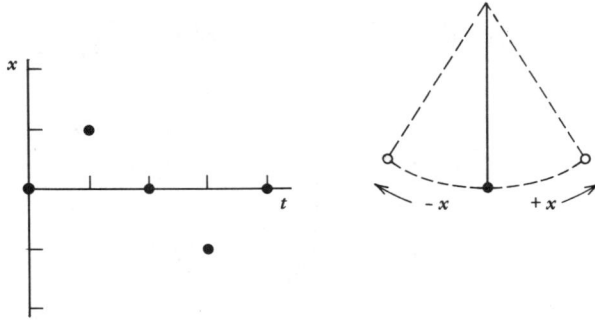

Fig. 15-1

the graph. After the same time has elapsed again, the bob is at its furthest extension to the left, where we have recorded its position as maximum negative. Another unit of time, and the bob is once again sweeping back through zero. We now have five points on a graph of position as a function of time. Without making any more observations we could guess about the function curve that ought to join those data points.

Why not join the data points with straight lines, as shown in Fig. 15-2? If that's the true function, consider the implications for the velocity. Plot the velocity as a function of time on the blank graph directly below the straight line graph.

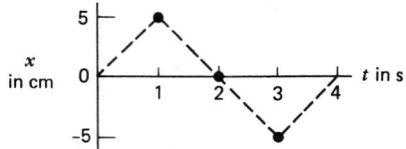

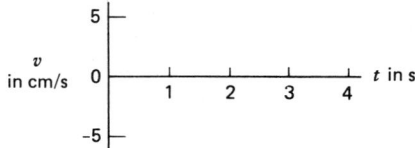

**Fig. 15-2**

**Question 15-1.** From your casual observations of the behavior of a pendulum, does the velocity of the bob follow the graph that you have drawn? What does your graph imply about the accelerations of the bob?

Since the bob has to move along a circular path, perhaps we should join the data points on the graph with semicircles, as shown in Fig. 15-3. Once again, consider the implications about the $v(t)$ curve for such a situation. What would be the velocity at the turnaround points? _____. What would be the velocity as the bob sweeps through the zero position? _____.

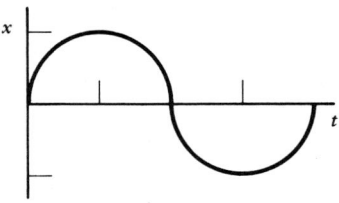

**Fig. 15-3**

Evidently, to describe cyclic phenomena, we need a mathematical function that crosses the horizontal axis *obliquely* and that turns over gradually at its maximum and minimum points. The sine and cosine functions have these properties. Take a look at the graphs you plotted on page 50.

Note how the graph of sin $\theta$ versus $\theta$ makes a plausible model for the motion of a pendulum bob. We were looking for a function which cuts the axis at an oblique angle and makes a smooth turnaround at the maximum. The sine function seems to fulfill those requirements.

We continue plotting sin $\theta$ for angles larger than 90° by appealing to symmetry. The motion of the pendulum bob is, after all, symmetrical, and the mathematical function describing it should display the same characteristics. We have indicated this continuation of the function in Fig. 15-4. Check to make sure that the graph is indeed symmetrical with the data points that you placed for the first quadrant in your graph on page 50. For instance, what is your value for 60° and what is the graph's value for 120°? What is your value for 45° and what is the graph's value for 135°? Beyond 180° the pendulum bob is swinging to the left of the starting position. We have arbitrarily defined that region to be negative, and, as you can see, the sine function is also negative in that region. The sine function oscillates back and forth endlessly, just as does the pendulum.

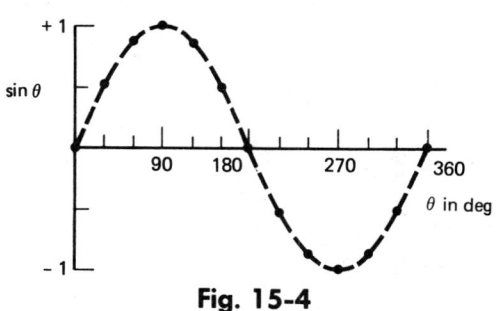

**Fig. 15-4**

Have we proved that the pendulum motion can be described by a sine function curve? Not at all. The function appears to be a plausible model for what we saw, but we would have to test its accuracy by actual measurements, or by showing analytically that the sine function ought to model a pendulum motion. As a matter of fact, the sine function is only an approximation to pendulum motion, good only for small angles of the swing amplitude.

## USES OF THE SINE FUNCTION

Many oscillation-type motions originate from *circular* motion. The tides, for instance, with water rising and falling vertically are produced by complicated motions of the earth, moon, and sun. However these motions are basically circular around each other. In a car engine, the pistons go up and down, but force the wheels to go in circles. In Fig. 15-5, we show a mounted wheel that is spinning with constant velocity. A ping pong ball is fastened to the rim and a spotlight is arranged to cast a shadow of this ball on a screen. The ball goes around and around in a circle, but the shadow oscillates up and down. Let's analyze the oscillating motion of the shadow in terms of the circular motion of the wheel. The position of the shadow above the zero position is $y = R \sin \theta$.

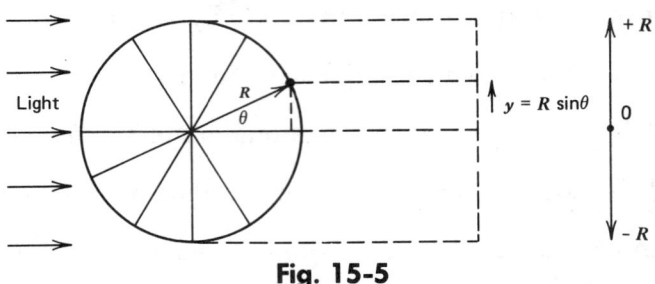

**Fig. 15-5**

**Question 15-2.** What is the displacement of the shadow from the zero position when $\theta = 0°$? 30°? 60°? 90°? 135°? 180°? 270°? 300°? 360°?

Since the wheel is traveling at constant velocity, we ought to be able to relate the angle $\theta$ to the time $t$. To do so, we must define angular, or rotational, velocity. Ordinary linear speed is linear distance traveled divided by the time of travel. Similarly, angular velocity is equal to the angular distance traveled divided by the time of travel. Angular velocity is given a special symbol $\omega$ (omega). $\omega = \frac{\Delta \theta}{\Delta t}$. In this form, the units are invariably radians per second (rad/s).

If at $t = 0$, we define the position of the wheel to be at $\theta = 0$, and if the wheel moves with constant angular velocity, then $\omega = \theta/t$, and $\theta = \omega t$. Instead of recording the position of the wheel in radians or degrees, we can simply time the movement of the wheel. The position of the shadow on the wall can now be described as $y = R \sin \theta = R \sin \omega t$. Time is not a cyclical variable; it goes on and on forever. However $\omega t$ represents an angle that is always a fraction or multiple of $2\pi$ radians, or 360°. As time goes on, as the angle gets larger and larger, the sine of the angle oscillates back and forth between $+1$ and $-1$.

The variable on which the sine function acts is called the *argument* of the function. The argument of any sinusoidal function must be an angle having no physical dimensions. The argument $\omega t$ contains time, but it also contains angular velocity with the dimensions, $1/T$ (radians *per second*). It is often convenient to use other arguments of the sinusoidal functions. Instead of measuring angular velocity in radians per second, we can also measure the revolutions, or cycles, per second. The symbol for cycles per second is $f$, and the internationally accepted unit is the hertz (Hz). As a wheel goes through one revolution, it goes through $2\pi$ radians. If a wheel is revolving at 10 Hz, then its angular frequency $\omega$ is equal to $20\pi$ rad/s. In general, we have $\omega = 2\pi f$.

There is another way to express the repetitive, or rotational, speed of a system; we can give the period $T$, the time it takes for one complete cycle.

**Question 15-3.** If $f = 10$ Hz, the system goes through ten oscillations, or revolutions, per second. What is the period? Therefore, what is the relationship between $f$ and $T$?

We can now write expressions for the position of the shadow of the ball on the rotating wheel using many different arguments for the sine function.

$$y = R \sin \theta = R \sin \omega t = R \sin 2\pi f t = R \sin 2\pi \frac{t}{T} = R \sin 360° \frac{t}{T}$$

Note that all these arguments are dimensionless. With the last two expressions, it's particularly easy to see how a repetitive function can be obtained from a variable, such as the time, that is increasing without limit. The sinusoidal function is being taken of a multiple of $2\pi$ rad, or $360°$. The multiple is $t/T$.

**Handling the Phenomena** Now that we have a mathematical function which appears to be a reasonable model for many oscillatory motions, we should test the agreement experimentally. In many cases, that's hard to do. For instance, we have suggested that the function represents the motion of a pendulum bob. Why shouldn't you now just hang a pendulum and measure the *position* of the bob as a function of *time* as it swings back and forth?

**Question 15-4.** Why would you have a hard time doing this with an ordinary stopwatch?

In the physics laboratory, you will have the opportunity to measure position as a function of time for objects that are undergoing very slow sinusoidal motion. Alternatively, you will be able to use fast

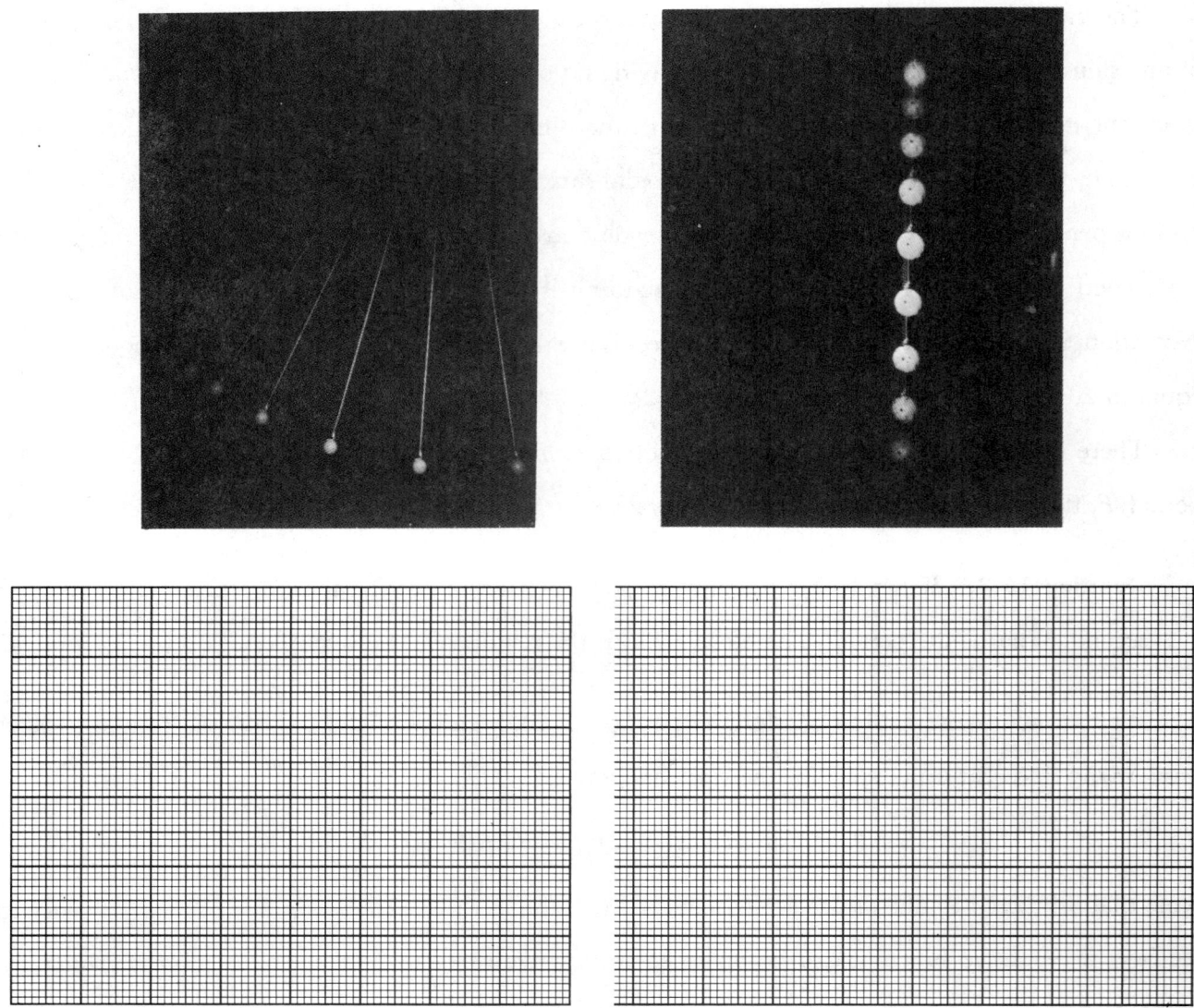

Fig. 15-6

timing methods to record rapid sinusoidal motion. We have provided two pieces of photographic data in Fig. 15-6. Each is a stroboscopic picture with the strobe light flashing at 30 times per second. The first shows a pendulum bob covering one quadrant of its path. The second shows a weight on a spring falling through $\frac{1}{2}$ of its path (from its largest positive extension to its largest negative extension). Use a ruler on each photograph to measure the positions of the bobs at each 1/30 s. Choose the center point (or equilibrium point) and measure the positions as distances from that center point. Plot the data on the graphs accompanying the pictures. In each case, draw the best smooth line through the data points.

Now check to see whether or not sine curves match the experimental curve. You will have to turn the time axis of each graph into an angle axis. For example, the time at the greatest extension of the bob corresponds to 90°. Plot values of $\sin \theta$ on each graph and see if they match the experimental data.

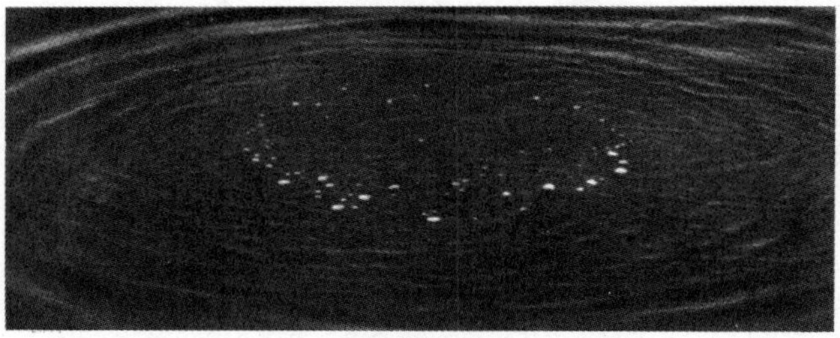

Fig. 15-7

## OTHER ARGUMENTS

So far, we have expressed the "argument" of a sine function in terms of an angle or some fraction of a complete period that depends on time. There are sinusoidal patterns in space as well as in time. In Fig. 15-7 we show a picture of ripples on a pond. A cross section of the surface of the pond would look very much like the graph of a sine wave. The ripples, or waves, are moving, and eventually we must be able to describe that motion. First, however, let's write the equation for the sinusoidal waves as the camera sees them at a given instant. The amplitude $A$ is the maximum height to which the water rises, or the maximum depth. The height or depth at any other point along the wave is given by

$$y = A \sin (\text{some function of } x)$$

The "argument" in parentheses must be some multiple of an angle, and of course must be free from any dimensions. Note that the waves repeat themselves in unit lengths called the wavelength $\lambda$ (lambda). We can specify the periodic nature of the waves by giving fractions of the wavelengths $x/\lambda$. We can choose our coordinates so that $y = 0$ when $x = 0$. As $x$ progresses toward $\frac{1}{4}\lambda$, $y$ increases to $A$. When $x = \frac{1}{2}\lambda$, the level of the wave is back to zero, or normal. For $x$ between $\frac{1}{2}\lambda$ and $\lambda$, the water level is negative, reaching its maximum negative amplitude $-A$, when $x = \frac{3}{4}\lambda$. To provide the proper argument for the sinusoidal function, the fraction $x/\lambda$ must be multiplied by $2\pi$ rad, or $360°$. Then, as $x$ increases without limit, the function itself oscillates back and forth between $+A$ and $-A$. The equation describing the waves at a given instant is

$$y = A \sin\left(2\pi \frac{x}{\lambda}\right) = A \sin\left(360° \frac{x}{\lambda}\right)$$

Study the way this works in Fig. 15-8.

**Question 15-5.** As you can see in the photograph, or as you may remember from watching ripples in a pond, at a great distance from the origin, waves die away. How would this affect our formula?

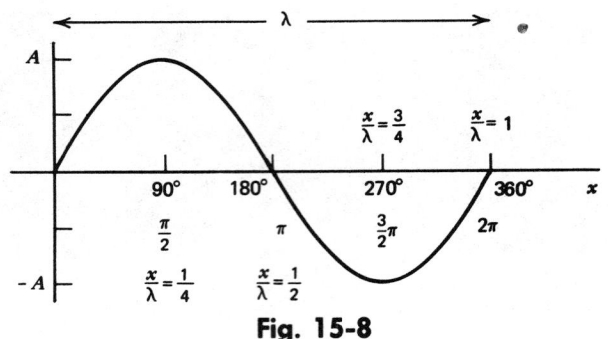

Fig. 15-8

The wave motion that we have described so far does not contain the variable of time. The formula describes only the height of the water as a function of $x$ at a given instant. If we watch the water at a particular point, then we see the height rise up and go back down again in a sinusoidal pattern that depends on time. Let's combine the time variable and the distance variable into one equation. Suppose at $x = 0$ we start our stop watch so that $t = 0$ just as a passing wave has $y = 0$. If we look over to our right at that instant, we will see the level of the water get higher and higher until at $\frac{1}{4}\lambda$ it has risen to a height of $+A$. On the other hand, if we keep watching the position $x = 0$, we will see the water there rise rapidly until when $t = \frac{1}{4}T$, the water there has risen to $+A$. We can combine both of these observations with the following formula

$$y = A \sin 2\pi \left( \frac{t}{T} - \frac{x}{\lambda} \right)$$

As time increases, the argument in the parentheses will get larger. On the other hand, as $x$ increases, the argument gets smaller. Suppose that you want to ride on top of the crest of the wave, maintaining $y = +A$. Time is increasing. In what direction are you and the wave heading if the sine function maintains its maximum value? Are you heading in the $+x$ direction or the $-x$ direction? _____ .

If you are going to keep up with the wave and maintain yourself on the crest, you must travel with the same speed as the wave. There is a relationship between the velocity, or speed, of the wave, the frequency, and the wavelength. As you can see in Fig. 15-9, if $f$ crests per second cross the observation line, and each of them represents a wave with a length $\lambda$, then the velocity of the waves must be

$$v = \lambda f = \frac{\lambda}{T}$$

# THE PHASE OF A CYCLIC ACTION

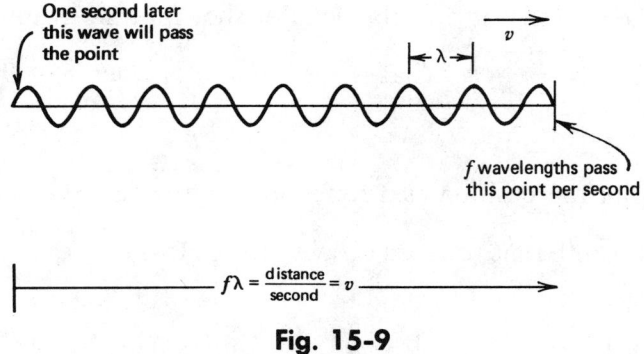

**Fig. 15-9**

**Question 15-6.** Check the units and dimensions of these relationships concerning $f$, $v$, and $\lambda$.

We can transform the sinusoidal wave equation so that the velocity of the wave becomes apparent.

$$y = A \sin 2\pi \left( \frac{t}{T} - \frac{x}{\lambda} \right) = A \sin 2\pi \left( ft - \frac{x}{v/f} \right) = A \sin \omega \left( t - \frac{x}{v} \right)$$

So far we have been talking about a wave moving to the right, in the direction of positive $x$. How could we change the wave equation to describe a wave moving to the left? ▢.

## THE PHASE OF A CYCLIC ACTION

For convenience's sake in describing cyclic action, we have usually assumed that $y = 0$ at $t = 0$ and also at $x = 0$. Of course, we might have chosen our origin at any other place, and at any other time. Look at the sine wave drawn in Fig. 15-10. At $\theta = 0$, $y$ is already up to one-half its final amplitude. Evidently, our origin is off by 30°. We can still describe the sinusoidal pattern by adding a *phase* angle.

$$y = A \sin(\theta + 30°)$$

| $\theta$ | $y$ |
|---|---|
| 0 | $\frac{1}{2}A$ |
| 30° | 0.87A |
| 60° | A |
| 90° | 0.87A |
| 120° | $\frac{1}{2}A$ |
| 150° | 0 |
| 180° | $-\frac{1}{2}A$ |
| −30° | 0 |
| −60° | $-\frac{1}{2}A$ |
| 240 | $-A$ |
| 330 | 0 |
| 420 | $+A$ |

**Fig. 15-10**

Check this equation to make sure that it describes what is shown in the graph. For instance, what is the value of $y$ when $\theta = -30°$? [ ]. What is the value of $y$ when $\theta = 90°$? [ ]. For what value of $\theta$ does $y = +A$? [ ].

Suppose four people plot the position as a function of time for *the same* pendulum bob. However, each starts a stop watch at a different time, as shown in Fig. 15-11.

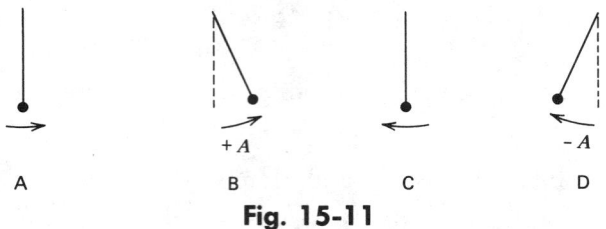

Fig. 15-11

Here are the graphs for the four observers. Part C has been drawn. You draw the other three.

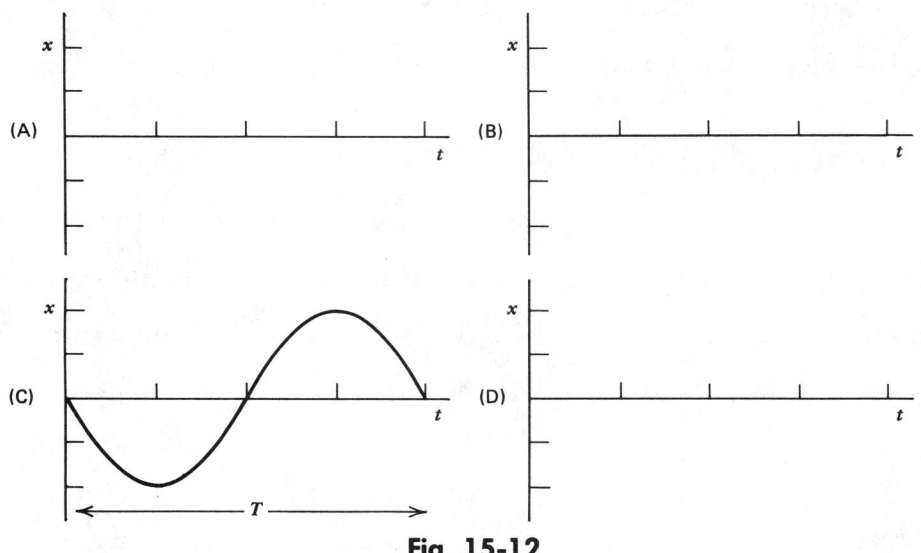

Fig. 15-12

The equation of the motion as measured by C is

$$x = A \sin\left(2\pi \frac{t}{T} + \pi\right) = A \sin\left(360° \frac{t}{T} + 180°\right)$$

What are the equations for $A$, $B$, and $D$?

_____

_____

_____

Note that in our descriptions of wave motion we could consider the *x* variable in the argument as providing a phase angle to the time-varying sine function. In this case, the phase varies depending on *x*.

**Question 15-7.** Suppose the phase angle is 90°: $y = A \sin(\theta + 90°)$. Sketch this function. What is its name?

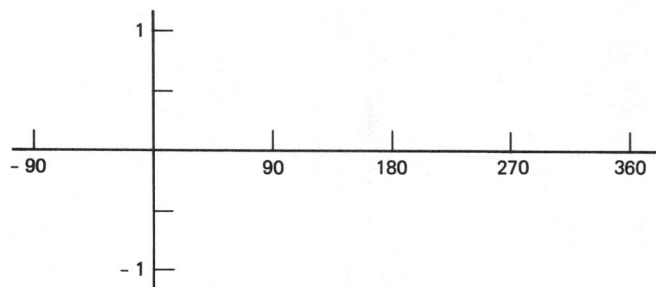

## THE SLOPES OF THE SINE FUNCTION

The way a function changes—the nature of its slope—is as important as the function itself. Remember that we started this section with sinusoidal functions by plotting $x(t)$ for a pendulum bob. Since at every point the slope of that graph gives the speed of the pendulum bob, we argued that the function itself must have certain properties. The turnaround at 90° had to be gradual, and the graph line had to cut the axis at an oblique angle so that the predicted speed would not be infinite. We then claimed that the sine curve had these properties and was a plausible description of $x(t)$ of a pendulum bob. Now let's take a closer look at the properties of the slopes of the sine curve.

In Fig. 15-13, we have three graphs. A sine curve has been carefully plotted in the upper one. With ruler and pencil, draw tangents to the curve at the points indicated and calculate the slopes of each of these tangents. Fill in the chart with your measured values. Because of the symmetry of the sine curve, it is necessary to make careful measurements only for the first 90°. You should be able to fill in the chart for the values greater than 90° without drawing tangents and making measurements.

**Question 15-8.** In measuring the slope of one of the tangent lines, you will be measuring the ratio $\frac{\Delta(\sin \theta)}{\Delta \theta}$. You can read the value for $\Delta(\sin \theta)$ directly off the left-hand axis. What units, however, should you use for $\Delta \theta$?

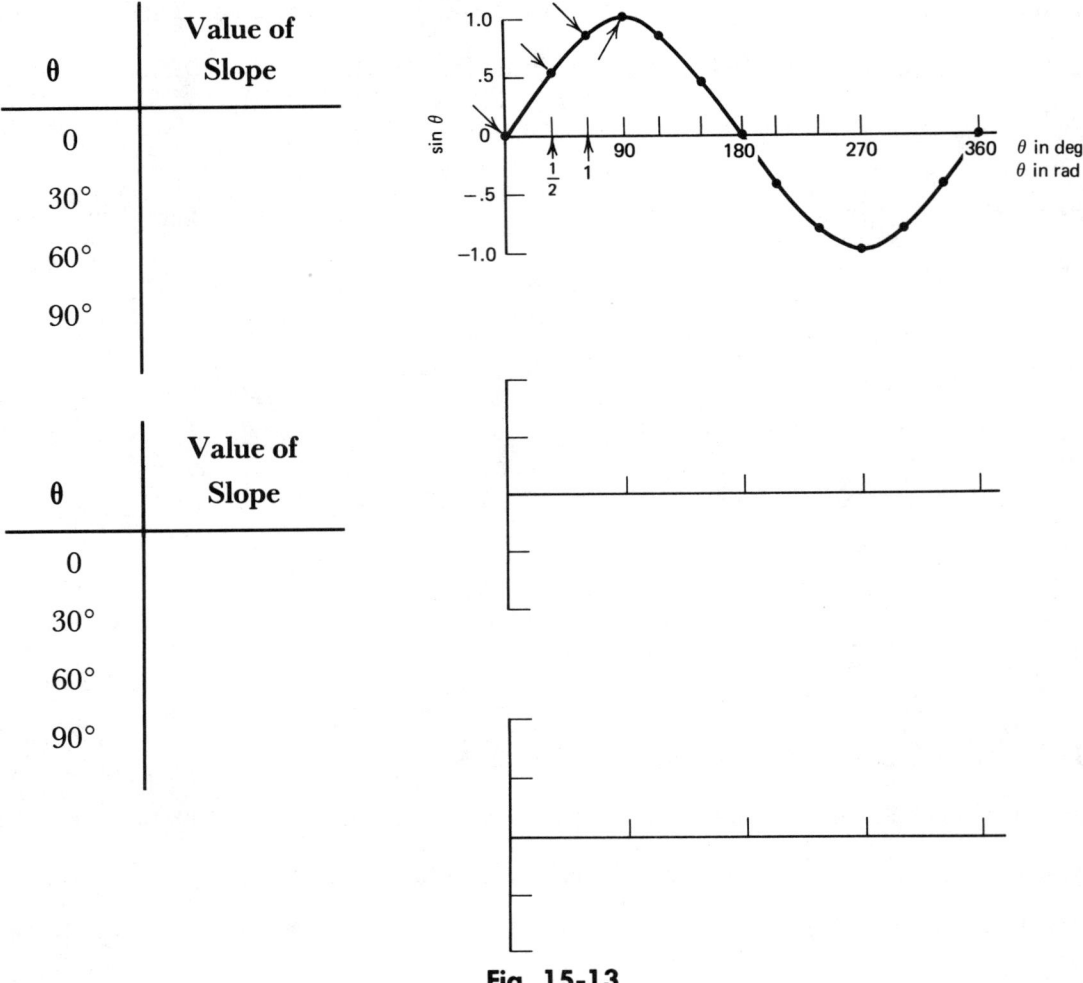

**Fig. 15-13**

After you have filled out the chart for the values of the slopes of sin $\theta$, enter the data points on the middle graph and connect the points with a smooth curve. What curve have you obtained? ▨ .

Now repeat the procedure of finding slopes for the same angles of the middle graph as you did for the sine curve. You will be finding the slopes of the slope function of the sine curve. Fill out the chart for this, enter the data points on the bottom graph, and draw a smooth curve through the points.

**Question 15-9.** How would you define the bottom graph that you have drawn?

Note the remarkable property of the sine curve in describing cyclic motion. First, it describes a repetitive motion of an object that goes back and forth between maximum and minimum limits.

Beyond that, however, the sine function also requires that the speed of the object is repetitive in the same way, speeding up and slowing down, reversing direction, speeding up in the opposite direction, and then slowing down again. Furthermore, the speed of the object is 90° *out of phase* with the position of the object. When the position of the object is at zero ($\sin 0 = 0$), the speed of the object is at its maximum ($\cos 0° = 1$). When the position of the object has reached its maximum ($\sin 90° = 1$), the speed of the object has slowed to zero ($\cos 90° = 0$). The mathematics describes exactly what happens with the pendulum bob. As the bob swings through its lowest point where $x = 0$, the speed is the greatest. When the pendulum bob reaches its maximum position, the speed has gone to zero and then will become negative as the bob goes back toward zero.

The slope of $v(t)$ is the acceleration. Therefore, the bottom graph that you drew represents the acceleration of an object that is moving with sinusoidal motion. It, too, is a repetitive function, and indeed it is simply a sine wave out of phase by 180°. When the pendulum bob is at its lowest point $x = 0$, the acceleration is also zero, although the speed at that instant is maximum. As the pendulum bob comes momentarily to rest at its maximum position, the acceleration is a maximum in the negative direction. The acceleration always has the opposite sign from the position. If the pendulum bob is in the region to the right, the acceleration is toward the left.

Compare the behavior of the slope function of the sine curve with that of the power functions. The slope functions of the power functions were also power functions, with the exponent reduced by 1. The slope function of the sine curve is also a sine curve, but moved in phase by 90°.

**Handling the Phenomena** The strobe photos of the swinging pendulum, or the bouncing spring bob, produced sinusoidal graphs of position as a function of time. The same pictures can be used to analyze the speed of the moving objects. Instead of plotting $x(t)$, plot $\frac{\Delta x}{\Delta t}$ versus $t$. You can get these measurements from the original photographs by measuring the distance between dots. The distance between dots is $\Delta x$ and since the interval between strobe lights was $\Delta t$, measuring $\Delta x$ gives you a quantity proportional to the speed during that time interval.

On the graph below, plot $\Delta x$ versus $t$ for the oscillating spring and compare the curve with the one that you obtained when you plotted $x(t)$.

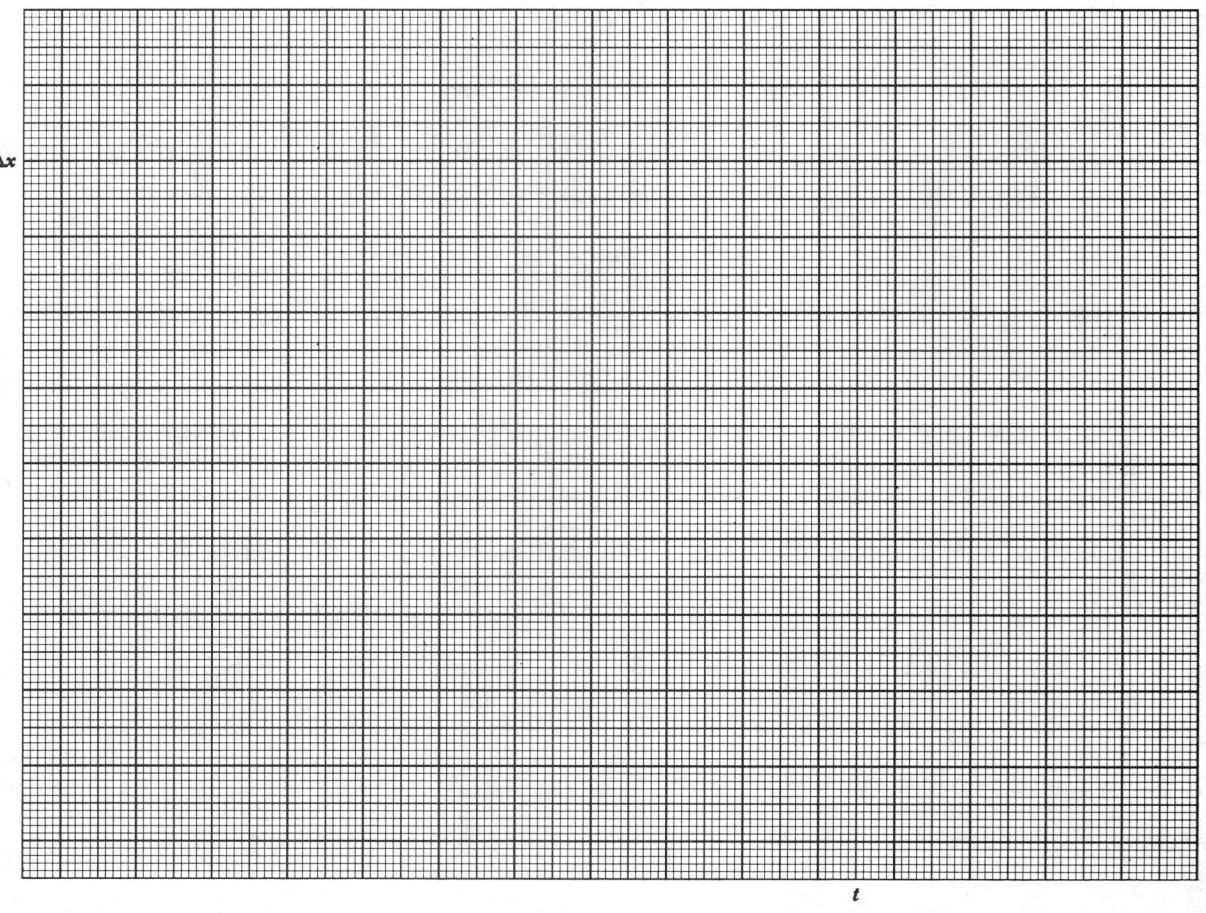

## Answers to Questions

15-1. The graph calls for constant speed of the pendulum bob—first to the right, then to the left. The turnaround time between positive speed and negative speed would be instantaneous, implying infinite acceleration. This situation seems hardly likely.

15-2.

| θ (deg) | Displacement | θ (deg) | Displacement |
|---|---|---|---|
| 0 | 0 | 180 | 0 |
| 30 | $\frac{1}{2}R$ | 270 | $-R$ |
| 60 | $0.87R$ | 300 | $-0.87R$ |
| 90 | $R$ | 360 | 0 |
| 135 | $0.7R$ | | |

15-3. If the system goes through 10 rev/s, each revolution must take $\frac{1}{10}$ s. $T = 1/f$.

15-4. From your previous work with pendulums, you know that the period of a 1-m pendulum is about 2 s. To measure $x(t)$ during one-quarter of the swing, you would have to make several measurements during $\frac{1}{2}$ s. If you can only make measurements to $\pm 0.2$ s, your measurements of $x(t)$ would not be very good.

15-5. The amplitude $A$ must decrease as the wave spreads out.

15-6. $v = \lambda f$. m/s = m(1/s). On both sides the units are meters per second.

$$\frac{L}{T} = L\frac{1}{T}$$

The dimensions check.

15-7. $\sin(\theta + 90°) = \cos\theta$

Your graph should look like a cosine plot, with $\cos 0° = 1$ and $\cos 90° = 0$.

15-8. Don't use degrees! $\Delta\theta$ should be measured in radians (rad).

15-9. It ought to look like an upside-down sine curve: $-\sin\theta$.

# CHAPTER 16
# THE EXPONENTIAL FUNCTION: $y = a^x$

So far, we have talked about phenomena where one variable increases or decreases as a function of some power of another variable. For instance, the distance that an object falls from rest is proportional to the square of the time. We have also found the mathematical descriptions for cyclic processes. The distance of a spring bob from its equilibrium position is $y = A \sin 2\pi \frac{t}{T}$. How fast the values of these functions change as their controlling variables change plays an important role in the usefulness of the functions as models for physical behavior. We characterized the changes in the functions in terms of the slopes of their graphs. For instance, the slope function of the sine curve is a cosine curve. Now we want to consider a class of phenomena by starting out with a requirement on the slope of the function, rather than on the function itself. What we will get is a mathematical function that describes growth or decay that feeds upon itself.

Suppose you put money in a savings account. The interest you get will depend on the amount of money that you invest and the time you leave it in. If $N$ represents the number of dollars in your account at any given time, then the increase in your dollars is $\Delta N$. $\Delta N$ is proportional to $N$ and is also proportional to $\Delta t$, the time interval during which the interest is being paid. If you invest your money two months, for instance, you will get roughly twice the interest that you would get in one month. Consequently,

$$\Delta N = rN\Delta t$$

The proportionality constant $r$ represents the interest rate. Its numerical value depends on the state of the economy and also on the units used to measure time. If the units are years, for instance, then $r$ might be equal to 0.06/yr. This would be the value for a 6% savings account.

Let's plot your money as a function of time. A graph of $N(t)$ is shown as a partial sketch in Fig. 16-1.

# THE EXPONENTIAL FUNCTION: $y = a^x$

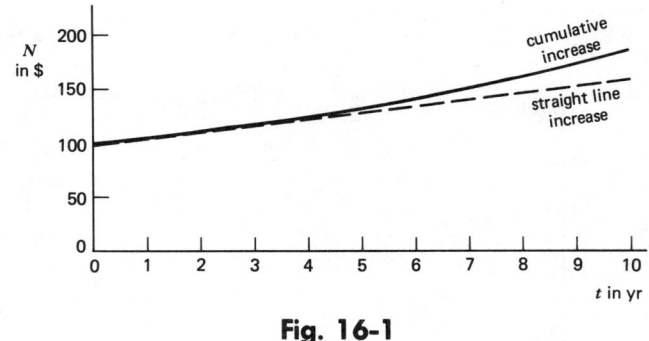

**Fig. 16-1**

**Question 16-1.** Why can't we draw the graph immediately as a straight line? If you have $100 to begin with, shouldn't you have $106 at the end of the first year, $112 at the end of the second year, and so forth?

On the sketched graph of $N(t)$, we could draw a tangent at a point and then complete the triangle to find the slope at that point. The slope is $\frac{\Delta N}{\Delta t}$. Our original formula for the growth of your money gives us an algebraic expression for that slope.

$$\frac{\Delta N}{\Delta t} = rN$$

Your money is represented by a function $N(t)$. We don't know what that function is yet, but we do know what its slope is. The slope of its graph at any particular time is proportional to the amount of money that is in the account at that particular time.

**Question 16-2.** Do any of the power functions have that property? What is the relationship of their slope functions to the original power function?

Of course, none of the sinusoidal functions have the desired property, either. The slope function for a sine curve is a cosine curve, which is a sine curve shifted in phase by 90°. Nevertheless, $\cos \theta$ is not proportional to $\sin \theta$. It looks as if we will have to invent a new function whose slope has the desired property.

**Handling the Phenomena** You can see for yourself what this new function must look like by sketching a curve on the graph below. In this case, for the sake of simplicity, $N$ and $t$ have the same units so that if $\frac{\Delta N}{\Delta t} = 1$, the slope will be 45°. Suppose at $t = 0$, we have $N = 1$. Let $r = \frac{1}{2}$. What is the slope then at $t = 0$? Draw a straight line with that slope from $t = 0$ to $t = 1$. What is the new value of $N$?

▢. What is the new value of $\frac{\Delta N}{\Delta t}$? ▢. Draw a straight line with that slope from $t = 1$ to $t = 2$. Continue on in the same way and observe how rapidly $N$ increases.

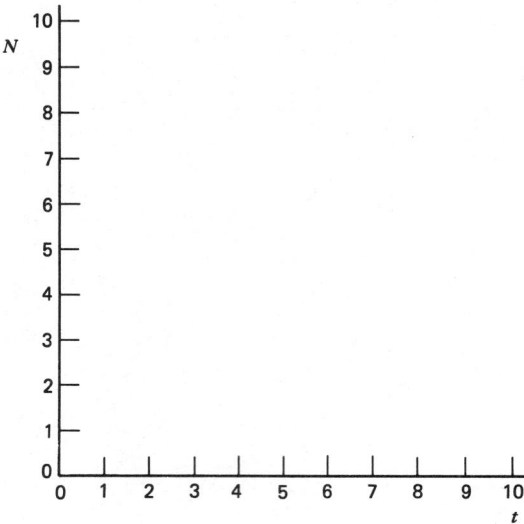

## THE BEHAVIOR OF THE EXPONENTIAL

Qualitatively, it appears that this new function grows very rapidly. It feeds on itself, so that the larger its value, the larger the slope of its graph. The larger the slope of its graph, the more the curve turns upward, and the larger the function becomes. These are the properties of the exponential function:

$$N(t) = N_0 a^{rt}$$

At $t = 0, N = N_0$.

Here's an algebraic proof that the exponential function has the desired slope properties. At one point in the derivation we must make use of a series substitution for $a^x$ when $x$ is small. For now, accept the substitution on faith. Before we begin the derivation, note that $a$ is not a variable. It's just a number like 2 or 10. With a power function, a variable is raised to a specific power. With the exponential function it's just the opposite. A number is raised to a power that is a continuously changing variable.

We assume that our function is

$$N(t) = a^{rt} \qquad (N_0 = 1)$$

At a time shortly later than $t$, the value of the function is

$$N(t + \Delta t) = a^{r(t+\Delta t)} = a^{rt} a^{r \Delta t}$$

(This last step is justified because $a^{x+y} = a^x a^y$. Remember that $10^{2+3} = 10^2 10^3$.) Now we will subtract $N(t)$ from $N(t + \Delta t)$.

$$N(t + \Delta t) - N(t) = \Delta N = a^{rt}a^{r\Delta t} - a^{rt} = a^{rt}(a^{r\Delta t} - 1)$$

Now we make use of the series substitutions. For small values of an exponent, we have

$$a^{r\Delta t} = 1 + kr\Delta t + \cdots$$

where $k$ is a proportionality constant that depends on the value of $a$. Since we are going to let $\Delta t$ go to zero, there is no point in continuing the series expansion further. Each successive term is smaller than the preceding one. After making the substitution, we get

$$\Delta N = ka^{rt}r\Delta t$$

Therefore, $\frac{\Delta N}{\Delta t} = rka^{rt} = rkN(t)$. The final equalities are true only in the limit as $\Delta t$ goes to zero. Under those circumstances, $\frac{\Delta N}{\Delta t}$ becomes the actual slope of the line at the point $t$. In calculus terms, $\frac{\Delta N}{\Delta t}$ becomes the derivative $\frac{dN}{dt}$.

The important point of the derivation is that we have found a function whose slope function is *proportional to itself*. Note that in the final step, $\frac{\Delta N}{\Delta t}$ is proportional to $a^{rt}$, which is just the function $N(t)$ with which we started.

**Handling the Phenomena**  Let us compare the properties of two functions that at first glance look very similar: $x^2$ and $2^x$. On page 164 there are tables of values to be filled out for these two functions and two blank graphs to be plotted. Note that at the beginning, $x^2$ rises more rapidly than $2^x$. However $2^x$ quickly takes over and soars out of sight.

## THE GROWTH POWER OF THE EXPONENTIAL

There is a story of an ancient kingdom where a wise man had done a service for a king. In the usual fashion, the king offered to grant the old man any gift up to one-half his kingdom. The sage said, "Sire, I am a humble man and do not desire much. Simply put one grain of rice on the first square of a chessboard, two on the next, four on the next, and so forth, doubling the number of grains for each remaining square of the chessboard."

**Question 16-3.**  On the first square of the chessboard there was $2^0 = 1$ grain of rice. On the next square there were $2^1 = 2$ grains of rice. There are 64 squares on a chessboard. How many grains were on the 64th square? How does this number compare with the number of grains on the 63rd square? How does this number compare with the number of grains on all the rest of the chessboard?

# THE EXPONENTIAL FUNCTION: $y = a^x$

| $x$ | $2^x$ |
|---|---|

| $x$ | $x^2$ |
|---|---|

**Handling the Phenomena** Take ten sheets of typing paper and number them from 0 to 9. Use shears or a paper cutter and cut paper number 1 exactly in half. Put paper 0, which had 0 cuts, at one end of a table and put beside it the two halves of paper number 1, piled on top of each other. Now take paper number 2 and cut it in half twice. That is to say, first cut it in half, then put those two pieces together and cut those in half again. You end up with four pieces of equal size. Put this pile of four pieces of paper neatly beside the first two. Continue on with the rest of the papers, using the number on the paper to tell you how many cuts you should make. Remember, each time you make a cut, pile the pieces neatly together so that the next cut will cut each and all of them in half.

Here is a variation on the paper cutting exercise. Get a large sheet of newspaper and fold it in half once. Then fold it in half again. Keep folding the resulting package in half and keep track of how many folds you make.

Note that in each of these cases the number of pieces of paper in the pile is equal to $2^x$, where $x$ is the number of times that you cut or fold. Meanwhile, what's happening to the size of each piece? If the original area is $A$, how would you define the areas of the resulting pieces? ▭ .

## THE BASE OF THE EXPONENTIAL

We started out by knowing something about the slope of a function, and now we have found a function that has that slope. Whenever we get a mathematical function we should plot it. The trouble is, we can't actually get numerical values for $a^x$ for various values of $x$ without knowing what $a$ is. The number $a$ is called the *base* of the exponential. It is not a variable; it can be any (positive) number that we choose. We have already seen a few examples of what happens if $a = 2$. You're probably also familiar with the use of the exponential function when $a = 10$. As you know, the use of powers of 10 in writing numbers is sometimes called "scientific notation."

Let's plot both $2^x$ and $10^x$ on the same graph. On page 166, there are two such graphs with different scales. Fill out the table of values for $2^x$ and $10^x$ and plot *both* curves on *both* graphs. As you will see, $10^x$ gets so large so rapidly as $x$ increases that it's hard to see the details of both functions on the same graph.

**Question 16-4.** For one value of $x$, the two exponentials have the same value. For what value of $x$ does this occur? Why does it occur?

On the second graph, which shows more details about the functions close to $x = 0$, draw tangents to the two curves at $x = 0$. What is the approximate value that you measure for the slope of $2^x$ at $x = 0$? _____. What is the approximate value that you measure for the slope of $10^x$ at $x = 0$? _____.

The slope that we derived for the exponential function is: $\frac{\Delta a^x}{\Delta x} = ka^x$. The proportionality constant $k$ depends on the base $a$. Since $a^0 = 1$ for any $a$, then you have just found with your measurements the values of $k$ for $a = 2$ and $a = 10$. You could probably measure those slopes only to

| $x$ | $2^x$ |
|---|---|
| $-1$ | |
| 0 | |
| 1 | |
| 2 | |
| 3 | |
| 4 | |

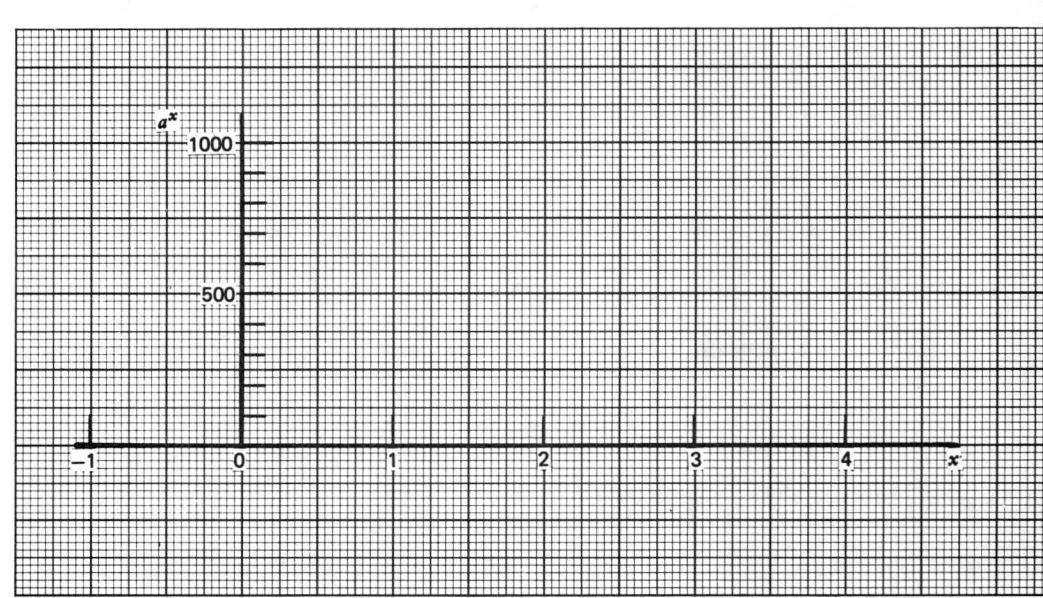

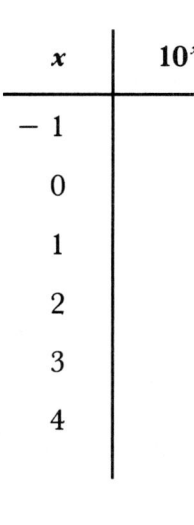

| $x$ | $10^x$ |
|---|---|
| $-1$ | |
| 0 | |
| 1 | |
| 2 | |
| 3 | |
| 4 | |

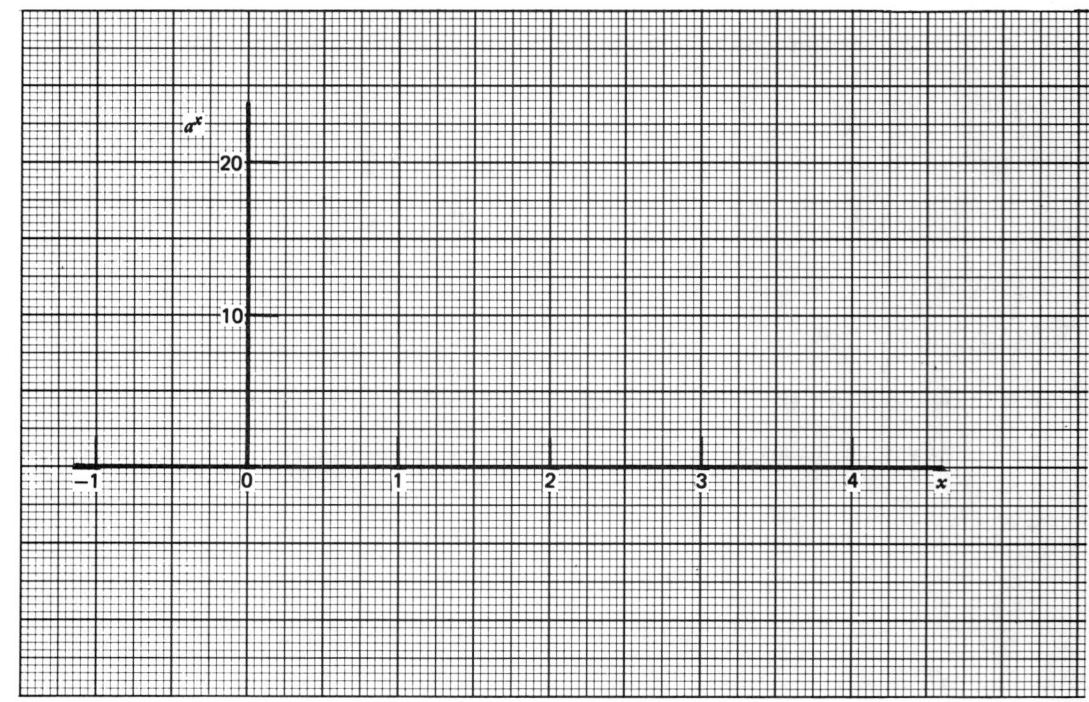

one significant figure. Here are the values of $k$ to two significant figures:

$$\frac{\Delta(2^x)}{\Delta x} = (0.69)2^x$$

$$\frac{\Delta(10^x)}{\Delta x} = (2.3)10^x$$

Is there some base of the exponential for which $k = 1$? If so, it must have a value between 2 and 10. That special number has the value 2.718.... It is called the *natural base* of the exponential function and is given the special name $e$. The special feature of $e$ is simply that $k = 1$ and, therefore, the value of the slope of $e^x$ for any $x$ is *equal* to the value of $e^x$ itself:

$$\frac{\Delta e^x}{\Delta x} = e^x$$

**Question 16-5.** The "natural" unit for measuring angles is the radian. What is the corresponding "natural" relationship between the sine function and its slope function if radians are used?

When we study the subject of logarithms, we will learn how to change bases of the exponentials and we'll also learn another way to graph them. In the meantime, here is a rule of thumb that we will prove later. If the annual percentage rate of increase of an exponentially growing function is $r$, then the doubling time in years is approximately $70/r$. For instance, if the inflation rate for the cost of living is 10% increase per year, then the doubling time for the cost of living is $(70/10) = 7$ yr. If you deposit money in a savings account at 5% compound interest per year, then the doubling time is $(70/5) = 14$ yr.

**Question 16-6.** Suppose that one of your ancestors had deposited $100 in a bank 280 years ago at 5% annual compound interest. If you now inherit this money and close out the account, how much do you get?

There are some useful approximations for transforming from $2^x$ and $e^x$ to $10^x$; for instance, $2^5 = 32$. Therefore, $2^{10} = 2^5 2^5 = 32 \times 32 = 1024 \approx 1000 = 10^3$. Higher powers of $2^x$ can then be broken down into multiples of $10^3$; for instance, $2^{25} = 2^{10} 2^{10} 2^5 \approx (10^3)(10^3)(32) \approx 3 \times 10^7$.

The easy relationship to remember for $e^x$ is that $e^3 = 20.09 \approx 20$. Then $e^{15} = (e^3)^5 \approx (20)^5 = (2)^5 \times 10^5 = 3.2 \times 10^6$.

## DECAYING EXPONENTIALS

So far, we have been dealing with exponential *growth*. As $x$ increases, $a^x$ rapidly gets larger. There can also be exponential *decay*. It occurs whenever the decrease in the original amount of something is proportional to the amount that already exists. An example is the way radioactive materials decay. If there are $N$ radioactive atoms, then the number that decay, $-\Delta N$, will surely be proportional to $N$. (The minus sign indicates a loss; $N$ is getting smaller.) You will get twice as many decays from 2 million atoms as you will from 1 million atoms. The number of atoms that decay in a short interval of time is also proportional to the time interval $\Delta t$.

$$-\Delta N = \lambda N \Delta t \quad (\lambda \text{ is the decay proportionality constant})$$

Let's assume that $N(t) = N_0 2^{-t/T}$. What does $N$ equal at $t = 0$? _____. What does $N$ equal when $t = T$? _____. It appears that $T$ is the "half-life." In a time $T$, one-half of the original atoms have decayed.

**Question 16-7.** Suppose for a particular radioactive material, $T = 1$ min. If a million atoms exist at $t = 0$, how many will be left at $t = 4$ min?

On the graph below, plot $N(t)$ from $t = 0$ to $t = 6$ min for $N_0 = 1000$, and $T = 1$ min.

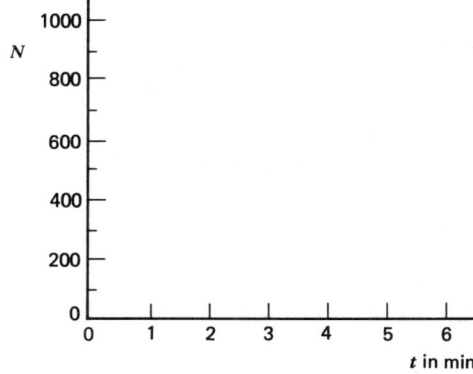

There are two special things to note about what we have just done. First, the exponent in the exponential function must be a number; it cannot have dimensions. You can not, for instance, raise 2 to a velocity power. Neither could we raise 2 to the time power with $t$ measured in minutes. Instead, we made the *argument* of the exponential function dimensionless by dividing $t$ by $T$. Both have the dimensions of time. The second point to note is that the negative exponential can be transformed into a positive exponential simply by placing it in the denominator. For instance, we have

$$N_0 2^{-t/T} = \frac{N_0}{2^{+t/T}}$$

The third point to note concerns the physics of the situation. We usually do not measure the number of radioactive atoms that exist at any given moment. Usually we are counting the *decays* $\Delta N$. However since $\Delta N \propto N$, we get exactly the same form of the decay curve with exactly the same half-life.

**Handling the Phenomena** While you may have difficulty getting instruments and materials to measure radioactive decay, anyone with a large number of dice can watch the dice "decay." If you can't get dice, use sugar cubes. Choose one of the numbers of the die, or mark one side of the cube, to signify a decay. Start out with at least 100 such cubes. Cast them, and remove all of them that have the chosen number on top. You would expect that *on the average* $\frac{1}{6}$ of the cubes would be removed (but not necessarily exactly $\frac{1}{6}$ at any particular cast). On the graph on the next page plot the number of cubes still in the game after each throw. Keep on casting the dice, each time removing the ones that have the chosen number or mark on top.

Your data probably look quite ragged. Since chance determines which side will come up on each cube, there will be fluctuations from the reduction factor that you might expect. The smaller the number of dice involved, the more important these fluctuations will become. Nevertheless, draw the best smooth curve that you can and then determine the "half-life" where the unit of time is in this case the "throw."

You can reexamine this experiment and your data after you have studied logarithms. In the meantime, see if you can find a relationship between your measured half-life and the "rule of seventy" that we described earlier.

| Throw | $N$ Remaining |
|-------|---------------|
| 0     | 100           |

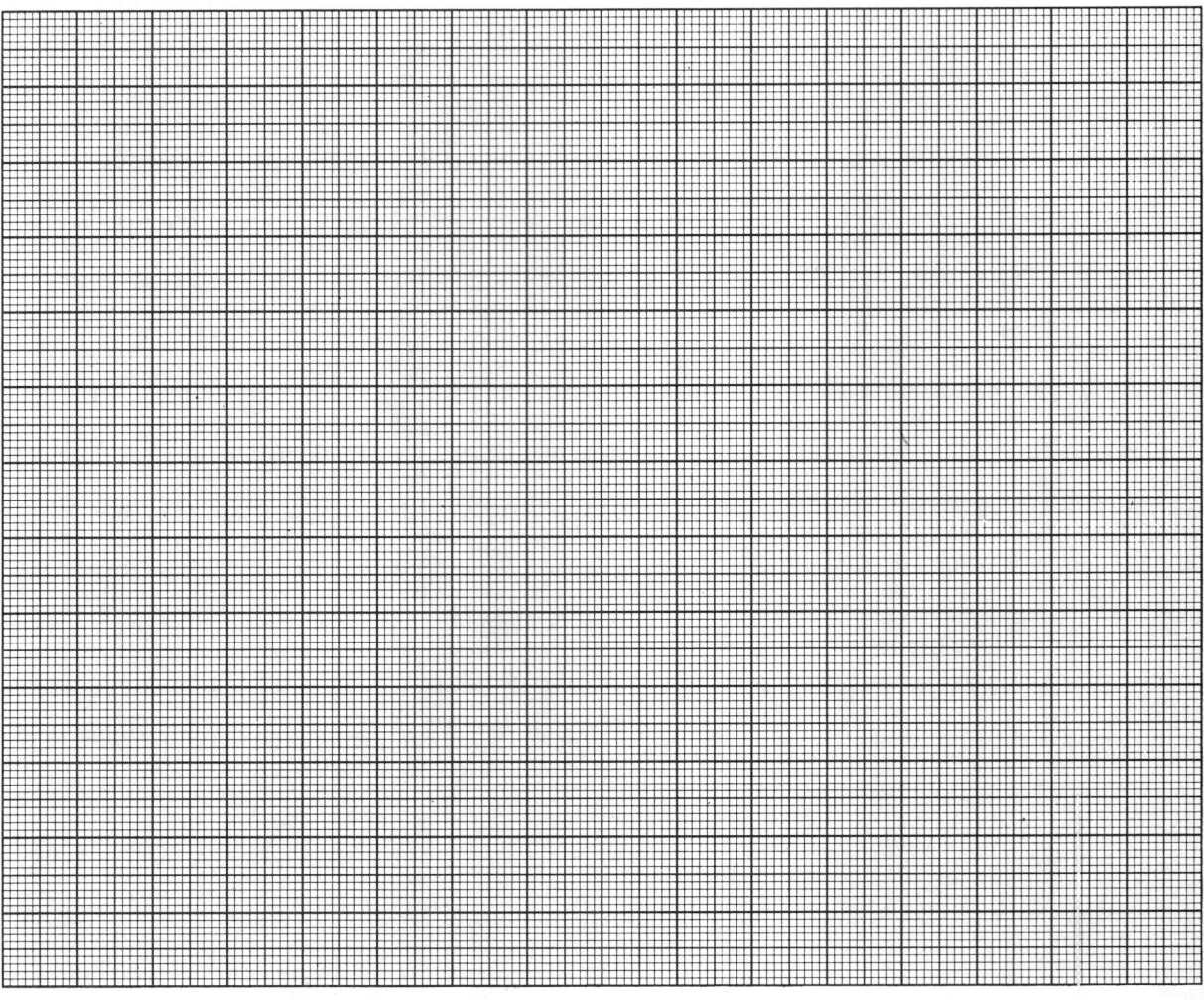

## Answers to Questions

16-1. If you have $106 at the end of the first year, you should get 6% interest on $106 during the second year. You should end up with $112.36 = $100(1 + 0.06)$^2$. After $x$ years you should have $100(1 + 0.06)$^x$.

16-2. The slope function (derivative) of the power function $y = kx^n$ is $nkx^{n-1}$. The slope function is not proportional to the original function but is proportional to a lower power of $x$.

16-3. On square 1 there was $2^0 = 1$ grain. On square 2 there were $2^1$ grains. On square $n$ there would be $2^{n-1}$ grains. Therefore on square 64 there would be $2^{63}$ grains. That is twice as many grains as there would be on the 63rd square. Note that on square 2 there is one more grain than on square 1. On square 3 there is one more grain ($2^{3-1} = 4$) than on the sum of squares 1 and 2

$(1 + 2 = 3)$. In general, on square $n$ there is one more grain than the sum on all preceding squares:

$$2^{n-1} = 1 + \sum_{x=1}^{n-1} 2^{x-1}$$

Therefore on the 64th square there is one more grain than on all the rest of the chessboard.

16-4. When $x = 0$, $a^x = 1$, regardless of the value of $a$. Remember that $a^x \times a^{-x} = a^x \times \dfrac{1}{a^x} = 1$. And $a^x \times a^{-x} = a^{x-x} = a^0$. So $a^0 = 1$.

16-5. The slope function (derivative) of $\sin \theta$ is equal to $\cos \theta$ if $\theta$ is measured in radians. Otherwise, the derivative of $\sin \theta$ is merely proportional to $\cos \theta$.

16-6. At 5% compound interest, the doubling time is about $70/5 = 14$ yr. During 280 years there would be $280/14 = 20$ doublings. Since $2^5 = 32$, $2^{10} = 32 \times 32 \approx 1000$. Furthermore, $2^{20} = 2^{10} 2^{10} \approx 10^3 10^3 = 10^6$. Your $100 would now be worth about $100 million.

16-7. $N = N_0 2^{-t/T} = 10^6 2^{-4/1} = \dfrac{1}{16} 10^6$.

# CHAPTER 17
# LOGARITHMS: $y = \log_a x$

Figure 17-1 shows a map of the universe. That may sound presumptuous. You know approximately how large a map of your state would have to be to show street details of your city. Of course, road maps are "linear" maps. If 1 in. represents 10 miles on the eastern border, 1 in. also represents 10 miles on the western border. As you can see, the map of the universe is not linear. It is *logarithmic*. Every division along the scale indicates an increase in size by a factor of 10. Note how the scale divisions go from $-15$ at one end to $+27$ at the other. Those division units actually stand for the exponents in powers of ten, and the units being used are meters. The scale mark 1 stands for $10^1$ m. The scale mark $-2$ stands for $10^{-2}$ m, or 1 cm.

**Question 17-1.** Is the position of the human reasonably indicated on the map? Shouldn't a human be taller than where we have placed him?

Look at the range of living creatures on this map of the universe. They occupy a region from about $10^{-7}$ m for the smallest viruses to only $10^2$ for the largest whales. Is there any limit to the sizes of the other objects? So far, measurements have not been made at distances smaller than $10^{-15}$ m, which is the size of a proton or a neutron in the atomic nucleus. At the other extreme, there appears to be an observational limit at $10^{26}$ m at the far edges of the universe. In our expanding universe, galaxies at that distance would be traveling close to the speed of light. Any signals from them would be Doppler shifted (red shifted) down into the low energy observational noise. Hence, they are forever beyond our observation.

There are many other types of phenomena whose response is most easily described by a logarithmic scale rather than a linear scale. As we shall see, most of our body senses are logarithmic detectors. In order to describe these processes, we will study logarithms as a mathematical function. In

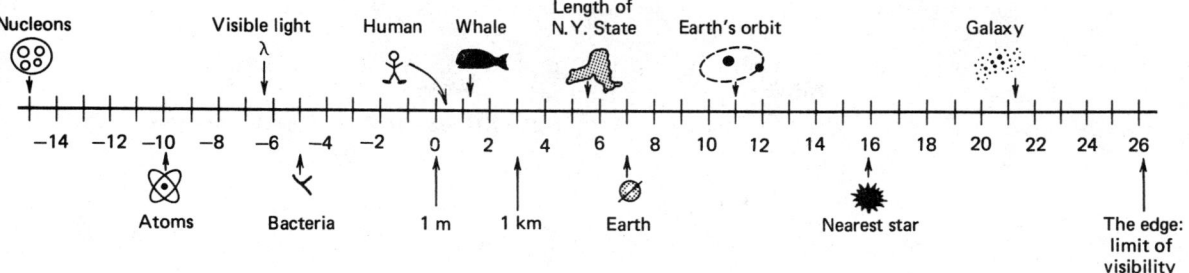

**Fig. 17-1**

the schools, logarithms are frequently taught, or at least introduced, as a method for performing long multiplication. They are rarely used for this purpose anymore, but the functional nature of the logarithm is crucial in our description of nature.

## CALCULATING LOGARITHMS

We have written below a number of values for the variable $x$. In the first row, the values are written in decimal notation. In the row directly underneath, the same values for $x$ are written again, but this time in the power-of-ten notation. In the third row, we make the same simplification that we did in labeling the scale of the map of the universe. We indicate only the exponent of the power of ten. Instead of calling it "exponent," we define this quantity with a new name: $\log_{10} x$ (log $x$ to the base 10). For instance, we could write $100 = 10^2 = 10^{\log_{10} 100}$, since $\log_{10} 100 = 2$.

| $x$ | 0.001 | 0.01 | 0.1 | 1 | 10 | 100 |
|---|---|---|---|---|---|---|
| $x$ | $10^{-3}$ | $10^{-2}$ | $10^{-1}$ | $10^0$ | $10^1$ | $10^2$ |
| $\log_{10} x$ | $-3$ | $-2$ | $-1$ | 0 | 1 | 2 |

**Question 17-2.** What is the value for $\log_{10} 1$? For $\log_{10} 0.001$?

We can operate with logarithms in the same way we do with exponents. For instance, when we *add* exponents, we perform *multiplication* of the variable $x$.

$$0.1 \times 100 = 10$$
$$10^{-1} \times 10^2 = 10^{2-1} = 10^1$$
$$\text{logs:} \quad -1 + 2 = 1$$

$$1000 \times 1000 = 1{,}000{,}000$$
$$(10^3)^2 = 10^3 \times 10^3 = 10^{3+3} = 10^6$$
$$\text{logs:} \quad 3^2 = 2 \times 3 = 6$$

# LOGARITHMS: $y = \log_a x$

For every value of $x$ (0.1, 10, 1000, etc.), we associate a value for $\log_{10} x$ ($-1$, 1, 3, etc.). Therefore, we have defined a *function* of $x$. Let's plot it. On the next graph we have labeled the axes for a plot of $\log_{10} x$ versus $x$. Enter the values that you know (and that will fit) and sketch a reasonable curve.

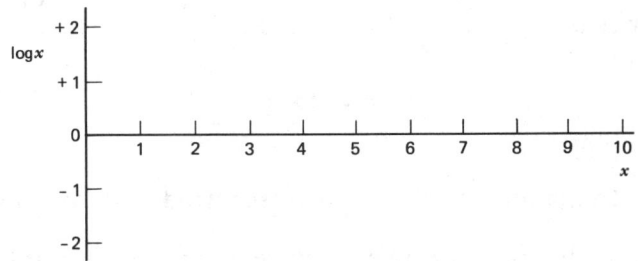

Note the nature of the logarithm function. A large change in $x$ makes only a small change in $\log x$.

**Question 17-3.** On the scale of this graph, how long would the $x$ axis have to be to show the point where $\log_{10} x = 2$?

So far it may seem that we have been playing games and making definitions about something that you already know how to do. You have probably been using power-of-ten notation for years. Without defining logs you can do the arithmetic of $10^2 \times 10^3 = 10^5$. You know that $10^2 = 100$ and $10^3 = 1000$. However, what good does it do to know that you can multiply powers-of-ten by adding their logarithms? Can you also find and deal with the logs of numbers like 3, 4, and 1848? Indeed, we can assign a log to every value of $x$, not just powers of ten. Let's do this first for the values of $x$ between 1 and 10. Follow the sample calculations below. These provide the values in the following table. Use the table values to complete your graph of $\log_{10} x$ as a function of $x$.

$$\sqrt[2]{10} = 10^{1/2} = 3.16$$

$$\sqrt[3]{10} = 10^{1/3} = 2.15$$

$$\sqrt[4]{10} = 10^{1/4} = 1.78$$

$$\sqrt[6]{10} = \sqrt[3]{\sqrt[2]{10}} = 10^{1/6} = 1.47$$

$$10^{5/6} = 10^{1/3} 10^{1/2} = 6.81$$

$$10^{2/3} = 10^{1/6} 10^{1/2} = 4.64$$

$$10^{11/12} = 10^{2/3} 10^{1/4} = 8.25$$

| $x$ | $\log_{10} x$ |
|---|---|
| 1.00 | 0 |
| 1.47 | 0.167 |
| 1.78 | 0.250 |
| 2.15 | 0.333 |
| 3.16 | 0.500 |
| 4.64 | 0.667 |
| 6.81 | 0.832 |
| 8.25 | 0.917 |
| 10.00 | 1.000 |

## MULTIPLICATION BY ADDITION

Now that we have logs for $x$ between 1 and 10, we could do simple arithmetic with them. For instance

$$\log_{10} 2.15 = 0.333 \quad \log_{10} 3.17 = 0.5$$

$$\log_{10}(2.15 \times 3.17) = \log_{10} 2.15 + \log_{10} 3.17 = 0.333 + 0.5 = 0.833$$

The number, whose log is 0.833, is 6.82, which is the product of 2.15 and 3.17. It appears that we can multiply two numbers by adding their logs, and then finding the *antilog* of the sum.

**Question 17-4.** Perform this operation for the product $2 \times 4$. Read the logs and antilog from the graph.

How can we use this method for values of $x$ larger or smaller than the range from 1 to 10? You already know how to express any number in terms of a number from 1 to 10 times a power of 10:

$$1470 = 1.470 \times 10^3 \quad 0.068 = 6.8 \times 10^{-2}$$

Since we know the logs of any power of 10 (just the exponent), and we have a graph of one decade of $x$ values, we can write the logarithm of any number:

$$\log_{10} 1470 = \log_{10} 1.470 + \log_{10}(10^3) = \log_{10} 1.470 + 3 = 0.167 + 3 = 3.167$$

$$\log_{10} 0.068 = \log_{10} 6.8 + \log_{10} 10^{-2} = 0.832 - 2 = -1.168$$

**Question 17-5.** Use log values from the graph to find the product of 230 and 6800.

Logs were invented in the early seventeenth century by a Scotsman, John Napier. They greatly facilitated calculations in navigation that required extensive long multiplication. To multiply two long numbers, it is faster to look up their logs in a table, add the logs, and find the antilog in the table, than to multiply the numbers directly. In our age of the calculator, no one has to use logs this way anymore.

Now we should extend the graph of log $x$ versus $x$ to negative values of the log. Note that the logarithm is defined only for positive values of $x$. For negative values of $x$, the exponent of 10 would have to be imaginary. In Fig. 17-2, we continue the plot of $\log_{10} x$ for values of $x$ between 0.1 and 1. Here is the way to find the value of $\log_{10} x$ for $x = 0.5$.

$$\log 0.5 = \log(5 \times 10^{-1}) = \log 5 + \log 10^{-1} = 0.7 - 1 = -0.3$$

| $x$ | $\log_{10} x$ |
|-----|---------------|
| 1.0 | 0 |
| 0.8 | −0.10 |
| 0.6 | −0.22 |
| 0.4 | −0.40 |
| 0.2 | −0.70 |
| 0.1 | −1.00 |

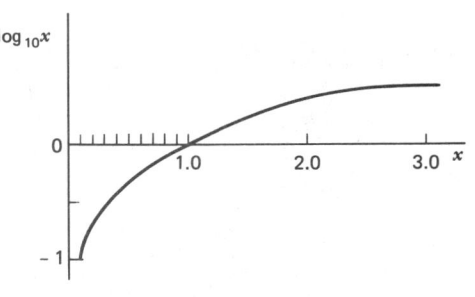

Fig. 17-2

## LOGS AND EXPONENTIALS TO OTHER BASES

So far we have dealt only with logs to the base 10. That is because we started with the familiar power-of-ten notation. If $x = 100 = 10^2$, then $\log_{10} 100 = 2$. It is frequently useful to express numbers in terms of powers of 2, or other numbers. For example, $32 = 2^5$. Therefore, $\log_2 32 = 5$. (Remember, the logarithm *is* the exponent.)

Let's work out some relationships that apply to logs to any base—even though in future technical work you will probably use only the base 10, perhaps base 2, and most of all base 2.72! (the *natural* base $e$ of the exponential). Call the base a particular number $a$. Then:

1. $(a^x)^2 = a^x a^x = a^{x+x} = a^{2x}$;

   for instance, $(2^4)^2 = 2^4 2^4 = 2^8 = 256 = 16 \times 16$

2. $a^x a^y = a^{x+y}$;

   for instance, $2^3 2^2 = 2^{3+2} = 2^5 = 32 = 8 \times 4$

3. $\log_a x^2 = 2 \log_a x$;

   for instance, $\log_2 4^2 = 2 \log_2 4 = 2 \times 2 = 4 = \log_2 16 = \log_2 2^4$

4. $\log_a(xy) = \log_a x + \log_a y$;

   for instance, $\log_2(4 \times 8) = \log_2 32 = \log_2 2^5 = 5 = \log_2 4 + \log_2 8 = \log_2 2^2 + \log_2 2^3 = 2 + 3 = 5$.

To change bases, we make use of the properties of both logarithms and exponentials, which, as you can see, are closely connected. We assert that we can change from one exponential base to another by multiplying the exponent by a constant:

$$a^x = b^{kx}$$

For instance, $4^x = 2^{2x}$. Take the logarithm to the new base $b$ of both sides:

$$x \log_b a = kx \log_b b = kx$$

since $\log_b b = 1$. Therefore, $k = \log_b a$.

$$a^x = b^{x \log_b a}$$

Let's apply this rule to find the relationships among the three common bases for exponentials:

$$e^x = 10^{x \log_{10} e} = 10^{0.434x} \qquad (\log_{10} 2.72 = 0.434)$$
$$10^x = e^{x \log_{10} e} = e^{2.3x} \qquad (2.72^{2.3} = 10)$$
$$2^x = e^{x \log_e 2} = e^{0.69x} \qquad (e^{0.69} = 2.72^{0.69} = 2)$$

Here is the method of changing bases of logarithms:

$$\text{if } y = \log_a x, \text{ then } x = a^y$$
$$\log_b x = y \log_b a = (\log_b a) \log_a x$$

Let's apply this rule to find the relationships between the two common bases for logs:

$$\log_e x = (\log_e 10) \log_{10} x = 2.30 \log_{10} x$$
$$\log_{10} x = (\log_{10} e) \log_e x = 0.434 \log_e x$$

## THE "NATURAL" LOGARITHM

The "natural" base of the exponential function is called $e$. It has a value of 2.718.... Because the exponential and log functions are so closely connected, $e$ is also the base of "natural" logarithms.

Instead of writing $\log_e x$, we usually just symbolize the natural log $x$ by writing $\ln x$. In the next section we will see what is natural about it.

**Question 17-6.** What is natural about $e^x$?

Here is a chart of powers of $e$, similar to the one we developed for powers of 10. Use the data to draw the graph of $\ln x$ from $\ln x = -2$ to $\ln x = +2$.

| $x$ | 0.14 | 0.37 | 1 | 1.65 | 2.72 | 7.39 |
|---|---|---|---|---|---|---|
| $x$ | $e^{-2}$ | $e^{-1}$ | $e^0$ | $\sqrt{e}$ | $e^1$ | $e^2$ |
| $\ln x$ | $-2$ | $-1$ | 0 | 0.5 | 1 | 2 |

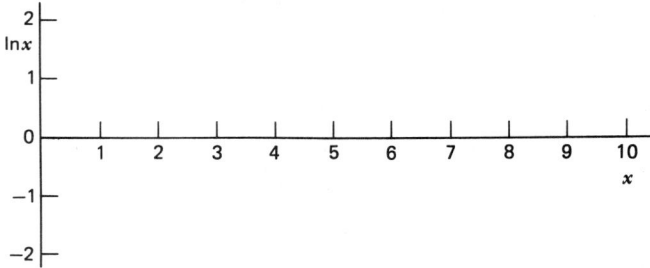

## COMPARISONS OF EXPONENTIALS AND LOGS

Figure 17-3 shows two graphs that illustrate the exponential and log functions for bases $e$ and 10.

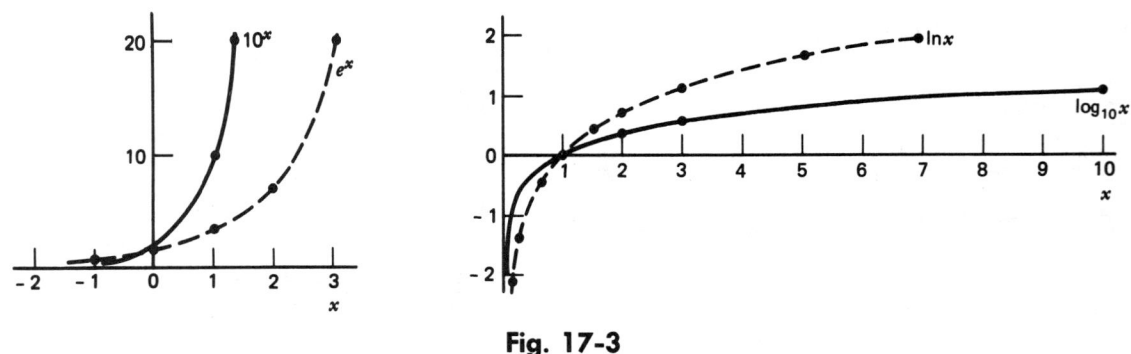

**Fig. 17-3**

Note that the value of the base affects the exponential through a multiplying factor *in* the exponential. (The scale of the horizontal axis changes.) The value of the base affects the log with a multiplying factor *in front* of the log. (The scale of the vertical axis changes.)

As we have seen, the logarithmic and exponential functions are intimately connected. The log of the exponential $a^x$ is the exponent itself. The relationship between log and exponential can be shown best on a single graph.

First, note that if $x = a^y$, then $y = \log_a x$. These are the *same* function. Plotting *either* one gives the graph shown in Fig. 17-4:

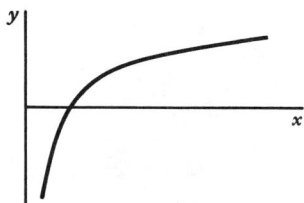

**Fig. 17-4**

If we *interchange* $x$ and $y$, we get

$$y = a^x \quad \text{and} \quad x = \log_a y$$

These two are the *same* function. Plotting *either* one gives the graph shown in Fig. 17-5.

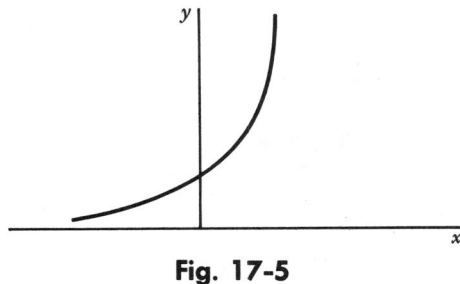

**Fig. 17-5**

Now we plot both the original function and the interchanged function on the same graph, as shown in Fig. 17-6.

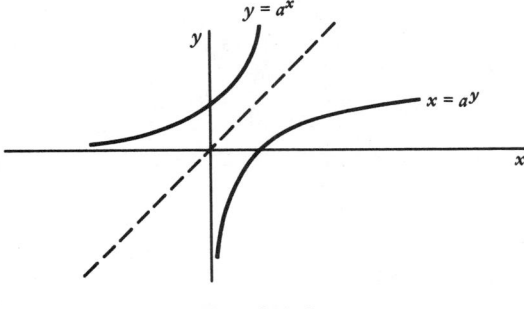

**Fig. 17-6**

Graphically, interchanging $x$ and $y$ corresponds to a reflection of the curve through the 45° line. Note that $y = a^0 = 1$ for any base $a$. Similarly, $y = \log_a 1 = 0$ for any base $a$.

**Question 17-7.** Describe in words the difficulty in plotting $e^x$ and $10^x$ on the same graph. Similarly, describe the difficulty of plotting $\ln x$ and $\log_{10} x$ on the same graph.

## THE SLOPE OF THE GRAPH OF THE LOGARITHM

With each of the functions that we have studied so far, the slope of the function graph has had special and useful characteristics. Let's find the "slope function" (in calculus terms, the derivative) of the logarithm. On the first graph on page 181, plot $\log_{10} x$ versus $x$ as accurately as you can, drawing a smooth curve through the points from $\log_{10} x = -1$ to $\log_{10} x = +1$. Then use a ruler and draw tangent slopes to the curve at five different points (perhaps where $\log_{10} x = -1, -0.5, 0, 0.5,$ and $1$). *Measure these slopes and plot the values on the second graph.*

The slope function that you get is a hyperbola with the formula $\dfrac{k}{x}$. The exact formula is

$$\text{slope function of log} = \left(\frac{d}{dx} \log_a x\right) = \frac{1}{x} \log_a e$$

If the base $a$ is 10, the slope function (derivative) is $\dfrac{1}{x} \log_{10} e = \dfrac{0.434}{x}$. If the base $a$ is $e$, the slope function is $\dfrac{1}{x} \log_e e = \dfrac{1}{x}$. Now we see what is "natural" about $\ln x$. Its derivative is *equal* to $\dfrac{1}{x}$, whereas for any other base the derivative is just proportional to $\dfrac{1}{x}$.

**Question 17-8.** What is the slope of $\log_{10} x$ for $x = 1$? What is the slope of $\ln x$ for $x = 1$? Compare your answers with the graph on page 181.

## SEMILOG GRAPH PAPER

In the chapter (Chapter 16) on the exponential function, you simulated radioactive decay by casting dice. The graph of the remaining number of dice $N$ as a function of the number of throws looked like an exponential decay curve. There is another way to plot an exponential function—one that yields a straight line. Suppose that the number of decays $\Delta N$ is proportional to the number $N$ and to the time interval $\Delta t$. Then we get $\Delta N = -\lambda N \Delta t$. The proportionality constant $\lambda$ (lambda), depends on the nature of the decay process (dice or radium). The minus sign indicates a decay process; for positive $N$, $\Delta N$ is negative. The slope of $N$ versus $t$ is $\dfrac{\Delta N}{\Delta t}$, which equals $-\lambda N$. The function that satisfies this slope condition is $N = N_0 e^{-\lambda t}$.

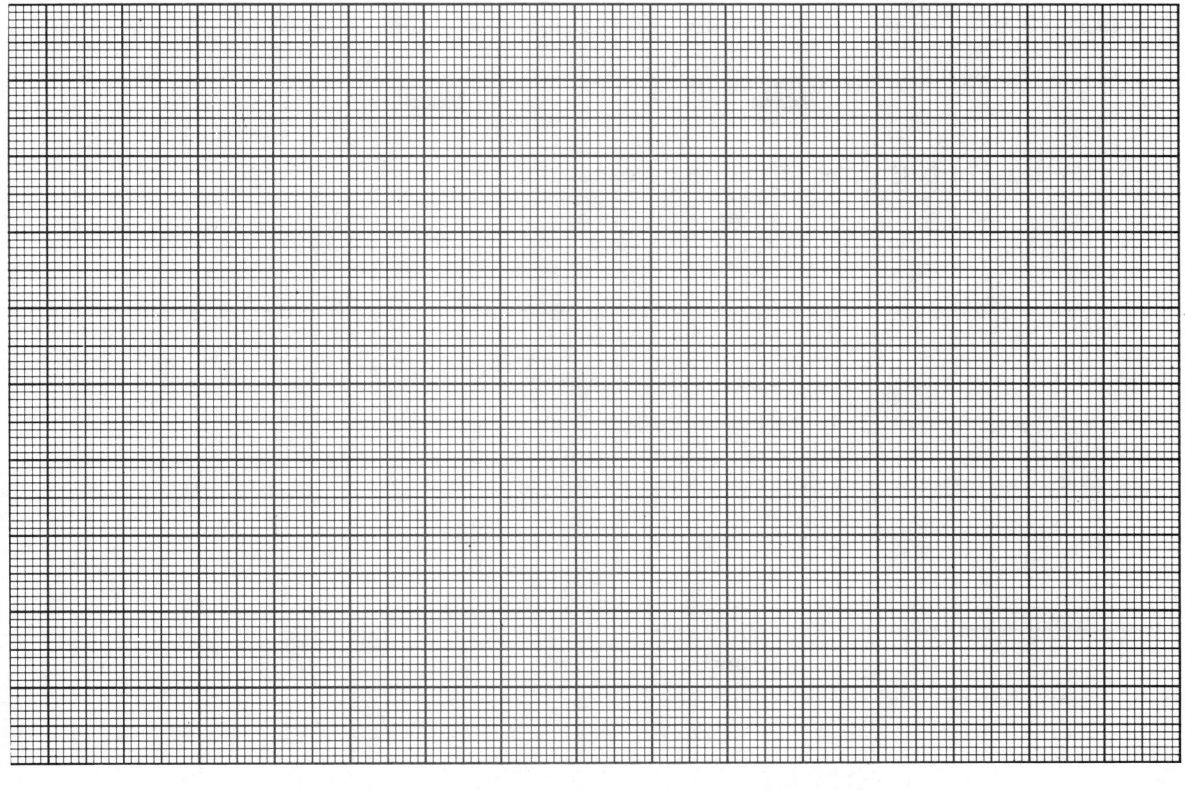
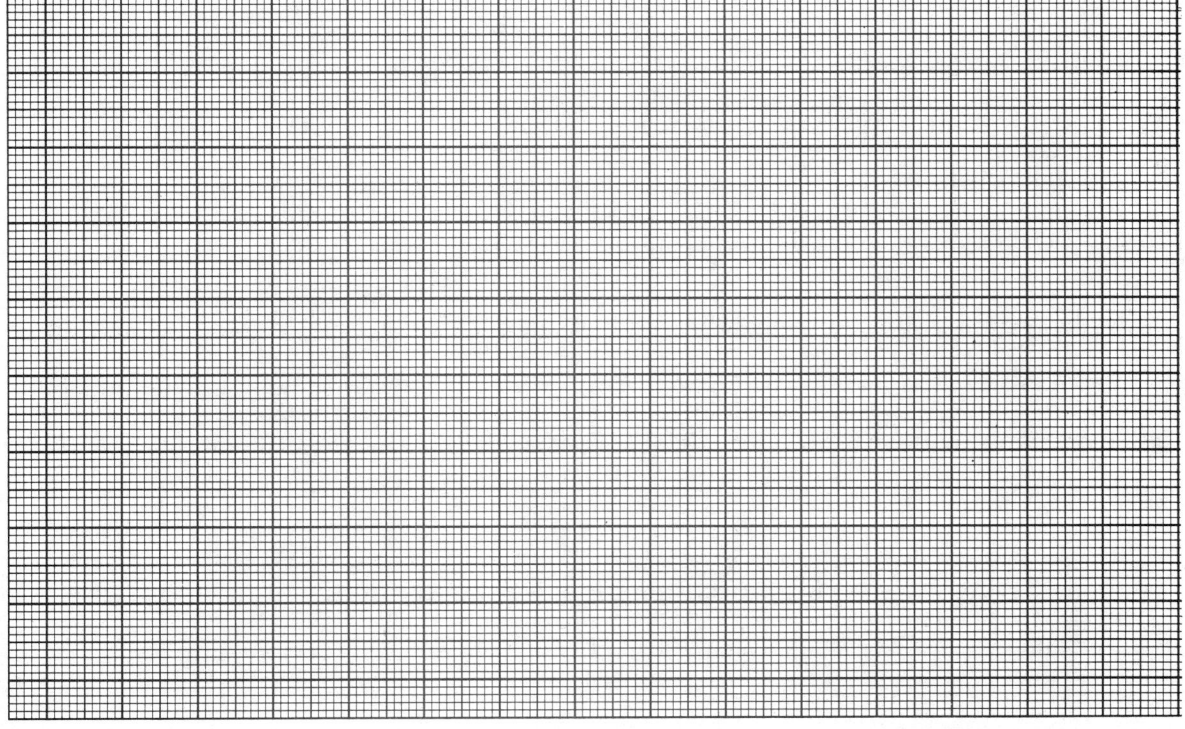

**Question 17-9.** For this equation, what is the value of $N$ when $t = 0$? What is the value of $N$ as $t$ goes to infinity? For what value of $t$ does $N = \frac{1}{2.7}N_0$?

Let's take $\log_{10}$ of both sides of the decay equation:

$$\log_{10} N = \log_{10} N_0 - \lambda t \log_{10} e = \log_{10} N_0 - (0.434\lambda)t$$

(since $\log_{10} e = 0.434$). If we consider $(\log N)$ the variable, instead of $N$, this is the equation of a straight line: $y = mx + b$. The intercept $b$ in this case is the constant $\log N_0$. One variable is $(\log N)$; the other is $t$. The slope $(m)$ of the line is $-0.434\lambda$.

Here are some data of a radioactive decay. Find $\log N$ and plot it against time on the first graph on page 183.

| $t$ (s) | $N$ | $\log N$ |
|---|---|---|
| 0 | $9800 \pm 100$ | |
| 10 | $6140 \pm 79$ | |
| 20 | $3840 \pm 63$ | |
| 30 | $2455 \pm 50$ | |
| 40 | $1540 \pm 39$ | |
| 50 | $1015 \pm 32$ | |

| $t$ (s) | $N$ | $\log N$ |
|---|---|---|
| 60 | $635 \pm 25$ | |
| 70 | $385 \pm 20$ | |
| 80 | $245 \pm 16$ | |
| 90 | $170 \pm 13$ | |
| 100 | $100 \pm 10$ | |

Study the second graph on page 183. The horizontal axis, for time, is standard and linear. The vertical axis, however, has its units plotted on a $\log_{10}$ scale. The distance between 100 and 1000 is the same as the distance between 1000 and 10,000. Furthermore, halfway between 100 and 1000 is not 550, but about 300.

**Question 17-10.** How could you construct graph paper like this? For instance, how could you find the proper distance between 100 and 1000 for the line corresponding to 300?

There is no zero on the log scale, since the decade below 100 would occupy an equal distance, ending at 10. Below this there would be a decade of equal size ending at 1. Below that, the decade would end at 0.1, etc. Graph paper of this kind, with one axis linear and one axis log, is called semilog. The example on the next page has two decades. You can get semilog paper with one or more decades,

SEMILOG GRAPH PAPER

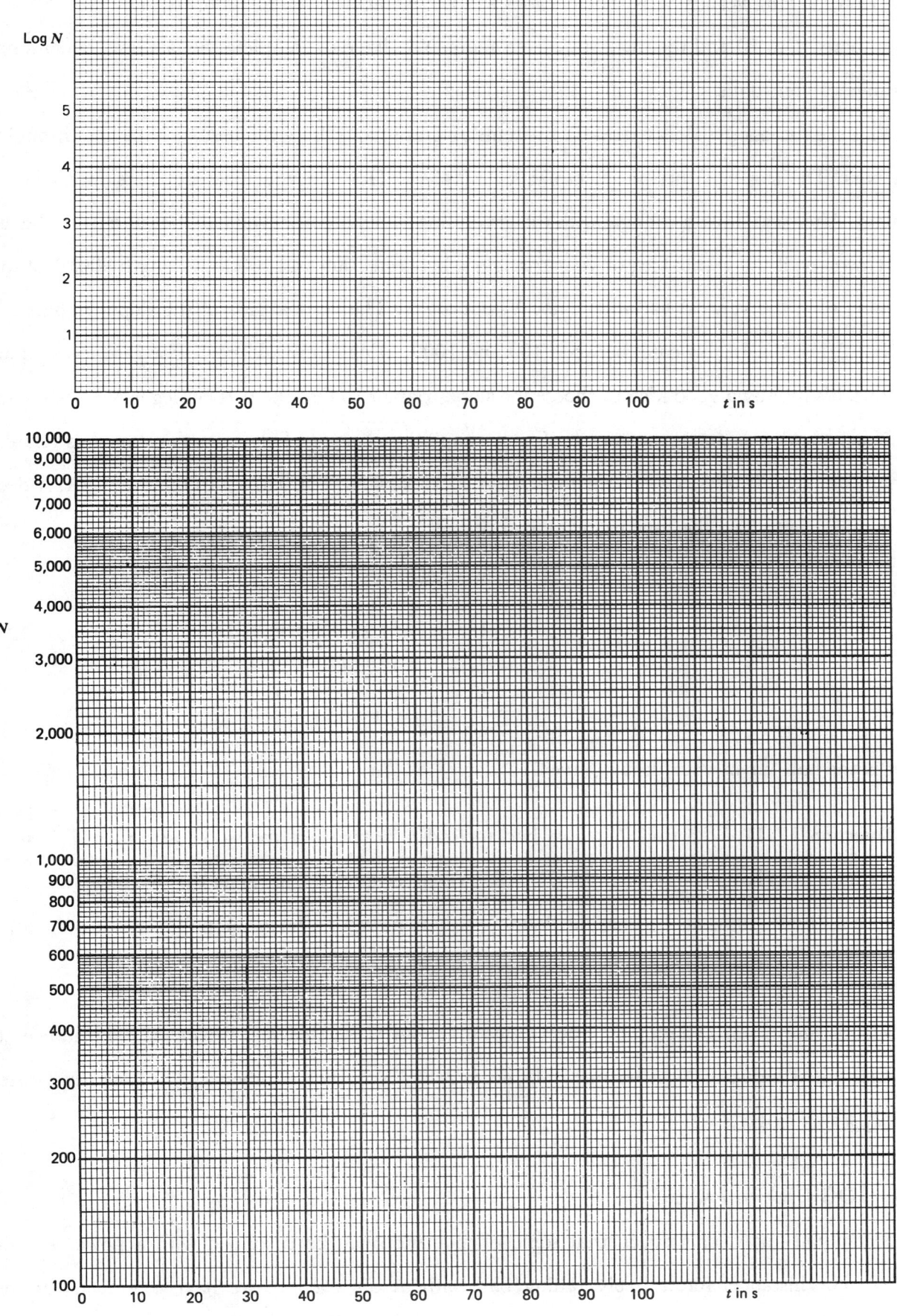

usually on the long vertical axis. The decades will not be labeled 100 to 1000, etc., however. Each will be numbered 1 to 10 and then will start over again. You have to complete the labeling, depending on the range of your data.

Plot the radioactive decay data on the semilog graph, and draw the best straight line through the data bars. The slope of this line is equal to $-0.434\lambda$. How do you measure the slope? There's a problem: After you draw a triangle with the graph line as hypotenuse, you can measure $\Delta t$ in the usual way. However, if the vertical leg of your triangle is between 1000 and 100, is its length 900? Not so! In a semilog plot, the vertical variable is not $N$, but $\log N$. The vertical side of the slope triangle is not $\Delta N$, but $\Delta(\log N)$. In this case, between 100 and 1000, $\Delta(\log N)$ would be 1. For ease in calculating the slope, it's usually a good idea to choose the slope-triangle so that the vertical leg is one decade long.

Now, what's the decay constant $\lambda$ of your graph? ☐ . The decay constant $\lambda$ is related to the half-life, $T_{1/2}$. If $N = N_0 e^{-\lambda t}$, then when $t = 1/\lambda$, $N = N_0/e \approx 0.37 N_0$. In the time $1/\lambda$, $N$ decays to $1/e$ of $N_0$. Let's find the time to decay by one-half:

$$\tfrac{1}{2} N_0 = N_0 e^{-\lambda t}$$

$$\ln(\tfrac{1}{2}) = -\lambda T_{1/2}$$

$$-0.69 = -\lambda T_{1/2}$$

$$T_{1/2} = 0.69/\lambda$$

What is the half-life of your decay curve? $T_{1/2} =$ ☐ s.

**Question 17-11.** To find $T_{1/2}$, we had to find $\lambda$ in terms of the slope, and then $T_{1/2}$ in terms of $\lambda$. What is the formula for $T_{1/2}$ directly in terms of slope? (Note that we have to deal with all these constants because the defining equation uses ln, the graph paper uses $\log_{10}$, and we want to know half-life, which is in terms of $\log_2$.)

Check both your linear and semilog graph to see that your calculated $T_{1/2}$ is correct. For instance, start with the largest value of $N$. After one half-life, is $N = \tfrac{1}{2} N_0$? After another $T_{1/2}$ is $N = \tfrac{1}{4} N_0$? After a third $T_{1/2}$ is $N = \tfrac{1}{8} N_0$?

## LOG – LOG GRAPH PAPER

Frequently you know, or suspect, that some phenomenon can be described with a power function: $y = kx^m$. For instance, for a freely falling ball without air friction, we have $y = \tfrac{1}{2} at^2$. Or for a pendulum, we have $T = (2\pi/\sqrt{g})L^{1/2}$. If you plot $y$ versus $t^2$ in the first case, or $T$ versus $L^{1/2}$ in the second case, you will get a straight line. What if you don't know the exponent of the power function? Or what if $m$ does not equal 2 or $\frac{1}{2}$, but is equal to 1.84? If you simply plot $y$ versus $x$, you will get a power function curve, but you won't know the power. Here's a way to find the exponent. Take the log of both sides:

$$y = kx^m$$

$$\log y = \log k + m \log x$$

This is the equation of a straight line if we let the variables be $(\log y)$ and $(\log x)$. The intercept is $\log k$, a constant, and the slope is equal to the exponent that we want to find.

Here is a table of values for $y = x^3$. Complete the columns by finding the logs of $y$ and $x$. Plot these values on the upper (linear) graph on page 186. Then use the lower graph to plot $y$ and $x$ directly. This lower graph is log–log. Both axes are calibrated with log scales. There are three decades each. Label these with appropriate units.

| $\log_{10} x$ | $x$ | $y$ | $\log_{10} y$ |
|---|---|---|---|
| | 0 | 0 | |
| | 1 | 1 | |
| | 2 | 8 | |
| | 3 | 27 | |
| | 4 | 64 | |
| | 5 | 125 | |
| | 6 | 216 | |
| | 10 | 1000 | |

The slope of the straight line on the log–log plot is equal to $\dfrac{\Delta(\log y)}{\Delta(\log x)}$. *If both axes have the same scale and same units*, the slope can be found with a ruler. Simply measure the vertical leg of the slope triangle with a ruler (in cm or any other units), and then measure the horizontal leg in the same units. The ratio of lengths is the slope.

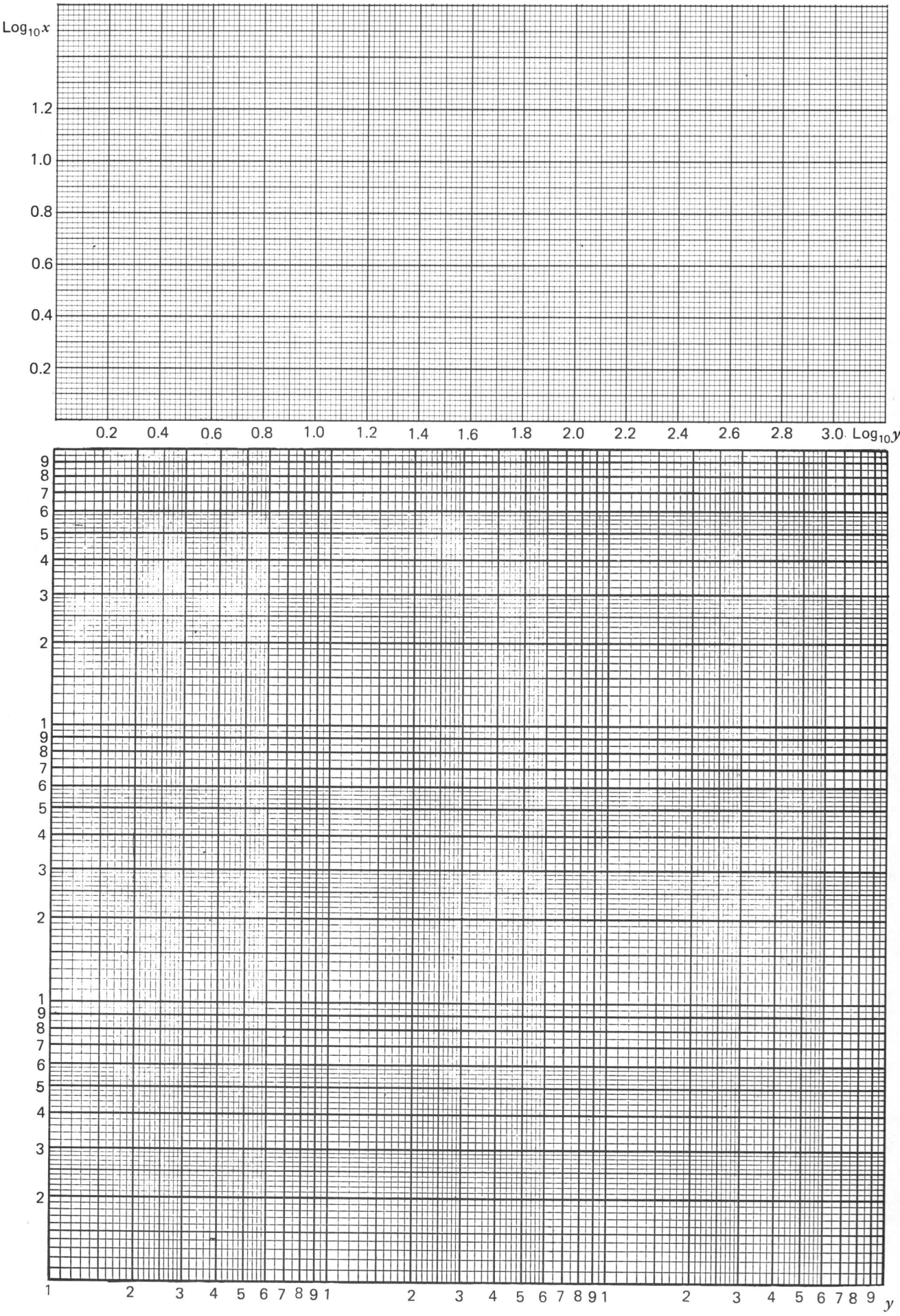

Check your graphs to make sure that the slope is 3. Here's a test of your understanding of the use of log–log graph paper. Find the algebraic relationship between $x$ and $y$ by plotting the following data on the log–log graph on this page.

| $x$ | $y$ |     | $x$ | $y$ |
|-----|------|-----|-----|-------|
| 0   | 0    |     | 4   | 12.00 |
| 1   | 1.50 |     | 5   | 16.80 |
| 2   | 4.25 |     | 6   | 22.10 |
| 3   | 7.81 |     |     |       |

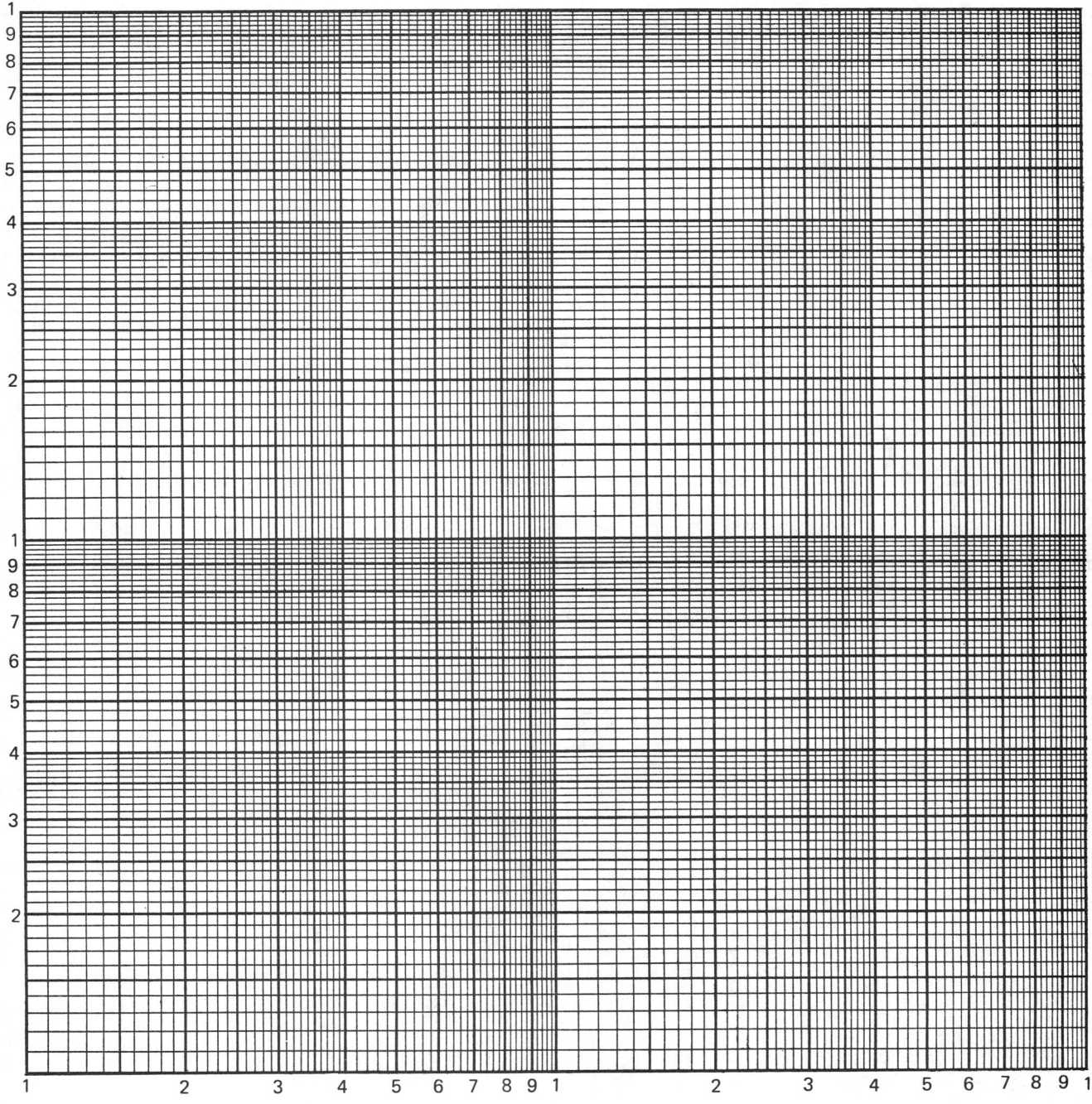

## USES OF THE LOG FUNCTION

### 1. The Measure of Acidity (pH)

Pure water is a good electrical insulator because it has very few free ions. The dissociation of $H_2O$ that does occur is into hydroxide ions ($OH^-$) and hydronium ions ($H_3O^+$ or larger clumps of $H_2O$ attached to an extra $H^+$). The concentration of these ions is usually described by chemists in terms of molarity, in this case the number of moles of ions per liter of solution. A mole (mol) is best thought of as a large number: 1 mol = $6 \times 10^{23}$. In pure water there are $10^{-7}$ mol of $H_3O^+$ ions per liter (L), or about 1 part in $5 \times 10^8$. ($10^{-7}$ mole is equal to $6 \times 10^{16}$. In 1 L of water there are 55.5 mol of $H_2O$ molecules, or $3.3 \times 10^{25}$ molecules.) There is an equal concentration of $OH^-$ ions.

It might be thought that if more hydronium ions were poured into the water, in the form of HCl for instance, the small quantity of $OH^-$ ions would all be neutralized, leaving no free hydroxide ions. Dissociation and recombination are always taking place, however, leading to a dynamic equilibrium. As the number of $H_3O^+$ increases by a factor of 10, the number of $OH^-$ decreases by a factor of 10, but does not go to zero. The product of the two concentrations remains constant:

$$K_{\text{water}} = [OH^-][H_3O^+] = 1 \times 10^{-14} \quad \text{at room temperature}$$

Since the ion concentrations vary over such a large range, the natural mathematical description is in terms of logarithms. The $p$H (potency of hydrogen) of a water solution is defined as the negative log of the molar concentration of hydronium ions. For pure water the $p$H is $+7$, since there are $10^{-7}$ mol of $H_3O^+$ per liter. The molar concentration of $H_3O^+$ in soda pop is $10^{-3}$; hence the $p$H is $+3$, and the molar concentration of $OH^-$ is $10^{-11}$. In 0.1 molar ($M$) sodium bicarbonate the $p$H is 8.4, therefore the molar concentration of $OH^-$ ions is $10^{-5.6}$. In this case the hydroxide ions dominate and the solution is basic, not acidic. The range of $p$H values for some common substances is

| | | | |
|---|---|---|---|
| 1.0 $M$ HCl | 0.1 | Rainwater | 6.2 |
| 0.1 $M$ HCl | 1.0 | Milk | 6.5 |
| Gastric fluid | 2.0 | Seawater | 8.5 |
| Lemon juice | 2.3 | Milk of magnesia | 10.5 |
| Vinegar | 2.8 | 0.1 $M$ NaOH | 13.0 |
| Tomatoes | 4.2 | 1.0 $M$ NaOH | 14.0 |

## 2. The Senses — Decibels

Most measuring instruments are sensitive only over a very narrow range. Voltmeters rarely have more than 100 divisions on their scale, and the fractional precision, of course, is inversely proportional to the magnitude of the reading. Even the standard meter stick has only 1000 scale divisions. As we move from one range to another of any variable that we are measuring, we must change instruments, or at least the scale of the instrument.

Fortunately for us, the senses of sight, feeling, etc., are exceptions to this rule. Although our precision is usually not great, still we must be able to sense forces from $10^{-3}$ to $10^{+3}$ $N$, from the weight of an insect to the weight of a man. We can see to some extent in starlight, when the illuminance is only $10^{-3}$ lumens per square meter ($lm/m^2$), and can also operate in bright sunlight at $10^5$ $lm/m^2$. Our sense of hearing covers a range of $10^{12}$. Furthermore, all these senses respond in a roughly logarithmic fashion. Not only do they cover a large range of magnitude, but their fractional precision remains about the same throughout the range. To illustrate this characteristic of logarithmic response, consider (or measure!) your sensitivity to differences in weights. Arrange the experiment so that weights are held by a thread in each hand and see whether you can detect a difference in weight if 1 g is held in one hand and 0.9 g in the other. When you are convinced that you can tell a difference of 0.1 g (0.001 $N$), or 0.2 g, see whether or not you can detect the same difference if one hand holds 1 kg and the other holds 999.9 or 999.8 g. Of course, you cannot. What you will find is that your *fractional* sensitivity remains about the same. If you can detect a difference of 1 part in 10 in one region of the range, that is approximately your sensitivity in any other region. The math model of this characteristic is that for some variable $x$ you perceive a response $y$, where $y = k \log x$. A change in the variable produces a change in the response as follows:

$$\Delta y = K \frac{1}{x} \Delta x \quad \left( \text{since} \quad \frac{d(\log x)}{dx} \propto \frac{1}{x} \right)$$

Equal increments of response $\Delta y$ correspond to equal *fractional* increments $\frac{\Delta x}{x}$ of the variable.

Corresponding roughly to human perception, the intensity of sound is measured on a logarithmic scale. In terms of intensity $I$, measured in watts per square centimeter ($W/cm^2$), the intensity level is defined to be

$$\text{decibels (dB)} = 10 \log_{10} \frac{I}{I_0}$$

On such a scale there must be a unit reference intensity. $I_0$ is usually chosen to be $10^{-16}$ $W/cm^2$, which is around the threshold of human hearing at 300 hertz (Hz) (cycles per second). The unit of intensity level is the decibel. Originally, intensity levels were defined in terms of $\log_{10}\frac{I}{I_0}$, and the unit was called the bel in honor of Alexander Graham Bell. The decibel is more convenient because the human ear can detect a difference in intensity level of about 1 dB. Note that if the actual energy intensity doubles, the ear hears an increase of 3 dB.

The intensity levels of various sounds are given below. They are only approximate because of the qualitative method of describing them and because the sense of hearing is strongly dependent on the frequency of the sound.

| Sound | Level (dB) | $I(W/cm^2)$ |
|---|---|---|
| Threshold | 0 | $10^{-16}$ |
| Whisper | 20 | $10^{-14}$ |
| Quiet office | 40 | $10^{-12}$ |
| Conversational speech | 60 | $10^{-10}$ |
| Heavy traffic | 70 | $10^{-9}$ |
| Passing subway | 90 | $10^{-7}$ |
| Hammering machinery | 110 | $10^{-5}$ |
| Pain threshold | 120 | $10^{-4}$ |

If the absolute intensity is measured in terms of the *pressure* of the sound, then the sound level in decibels is equal to $20\log\frac{p}{p_0}$, since the energy flow is proportional to the square of the pressure amplitude of the wave. There is a similar situation in electrical engineering measurements of electromagnetic energy flow. The decibel amplification is equal to $10\log\frac{E}{E_0}$ or $20\log\frac{V}{V_0}$, where $E$ is the measured energy and $V$ is the voltage.

## Answers to Questions

17-1. Most humans are a little shorter than 2 m. On the log scale, that height is closer to 0 ($10^0 = 1$ m) than to 1 ($10^1 = 10$ m).

17-2. $\log_{10} 1 = 0$, since $10^0 = 1$; $\log_{10} 0.001 = -3$, since $10^{-3} = 1/1000 = 0.001$.

17-3. The $x$ axis would have to be 10 times as long to get to $x = 100$. $\log_{10} 100 = 2$.

17-4. Find $\log_{10} 2$ on the graph that you have drawn ☐. Find $\log_{10} 4$ on the graph that you have drawn ☐. Add these two logs. ☐. What number has this log? ☐.

# USES OF THE LOG FUNCTION

17-5. $\log_{10} 230 = \log_{10}(2.3 \times 10^2) = \log_{10} 2.3 + \log_{10} 10^2 = 2 + \log_{10} 2.3 = \boxed{\phantom{xxx}}$. $\log_{10} 6800 = \log_{10}(6.8 \times 10^3) = \log_{10} 6.8 + \log_{10} 10^3 = 3 + \log_{10} 6.8 = \boxed{\phantom{xxx}}$. Add these two logs. Separate the decimal part and find its antilog. Your answer is this antilog time $10^6$.

17-6. The slope function (derivative) of $e^x$ is *equal* to $e^x$. The derivative of $a^x$ is only *proportional* to $a^x$.

17-7. On an ordinary graph if the vertical scale is chosen so that $e^x$ can be seen clearly for $x > 2$, the curve for $10^x$ will be off the top of the page. For the log functions, it is the horizontal axis that will not accomodate $\log_{10} x$ for $x > 1$.

17-8. Slope function of $\log_{10} x = \dfrac{1}{x} \log_{10} e = \dfrac{0.434}{x} \to 0.434$ for $x = 1$. Slope function of $\ln x = \dfrac{1}{x} \to 1$ for $x = 1$.

17-9. $N = N_0 e^{-\lambda t}$. As $t \to 0$, $N \to N_0$, since $e^0 = 1$. As $t \to \infty$, $N \to 0$, since $e^{-\infty} = \dfrac{1}{e^\infty} = 0$. If $t = \dfrac{1}{\lambda}$, $N = N_0 e^{-1} = \dfrac{1}{2.7} N_0$.

17-10. $\log_{10} 10 = 1$; $\log_{10} 100 = 2$; $\log_{10} 1000 = 3$. On a log scale the spacing between 10 and 100 is the same as the spacing between 100 and 1000. $\log_{10} 300 = 2 + \log_{10} 3 = 2.48$. The line corresponding to 300 is at 2.48 on the log scale.

17-11. $T_{1/2} = 0.69/\lambda$; slope $= -0.434\lambda$.

$$T_{1/2} = 0.69 \, (0.434/\text{slope}) = 0.3/\text{slope}$$

# APPENDIX 1
# USEFUL FORMULAS AND RELATIONSHIPS

## QUADRATIC

If $ax^2 + bx + c = 0$, then

$$x = \frac{-b \pm \sqrt{b^2 - 4ac}}{2a}$$

For instance,

$$3x^2 + 5x + 2 = 0$$

$$x = \frac{-5 \pm \sqrt{25 - 24}}{6} = -\frac{4}{6} \quad \text{or} \quad -1$$

## SQUARE ROOTS

The square root of 27 is $(5 + x)$, where $x \ll 5$:

$$27 = (5 + x)^2 = 25 + 10x + x^2$$

Since $x \ll 5$, $x^2 \ll 10x$ or 25 and can be ignored.

$$2 \approx 10x \qquad x \approx 0.2 \qquad \sqrt{27} \approx 5.2$$

This method can give you a square root to two significant figures, and can be done in your head. Try it with the following:

$$\sqrt{46} = 7 - x \qquad x = \boxed{\phantom{xxx}}$$

$$\sqrt{1840} = \sqrt{18.4} \times 10 = (4 + x) \times 10 \qquad x = \boxed{\phantom{xxx}}$$

## BINOMIAL EXPANSION

If $n$ is a positive integer, then

$$(a + b)^n = a^n + na^{n-1}b + \frac{n(n-1)}{2!}a^{n-2}b^2 + \frac{n(n-1)(n-2)}{3!}a^{n-3}b^3$$

$$+ \cdots + \frac{n!}{k!(n-k)!}a^{n-k}b^k + \cdots + b^n$$

$$(5! = 5 \times 4 \times 3 \times 2 \times 1)$$

For instance,
$$(a+b)^3 = a^3 + 3a^2b + 3ab^2 + b^3$$
If $n = -1$, $a = 1$, and $b = x$, where $x < 1$, then
$$\frac{1}{1-x} = 1 + x + x^2 + x^3 + \cdots$$

## APPROXIMATIONS

$$\sin\theta = \theta - \frac{\theta^3}{3!} + \frac{\theta^5}{5!} - \frac{\theta^7}{7!} + \cdots$$

$\sin\theta \approx \theta$     for small $\theta$     ($\theta$ in rad)

$$\cos\theta = 1 - \frac{\theta^2}{2!} + \frac{\theta^4}{4!} - \frac{\theta^6}{6!} + \cdots$$

$\cos\theta \approx 1$     for small $\theta$     ($\theta$ in rad)

$$e^x = 1 + x + \frac{x^2}{2!} + \frac{x^3}{3!} + \cdots$$

$e^x \approx 1 + x$     for $|x| < 1$

$2^5 = 32$    $\therefore 2^{10} \approx 10^3$

$e^3 \approx 20$

$\sqrt{1+x} \approx 1 + \frac{1}{2}x$     for $x$ small compared with 1

$\pi \approx 3$     $\pi^2 \approx 10$

## TRIGONOMETRY RELATIONSHIPS

$$\sin\theta = \frac{\text{opp}}{\text{hyp}} \qquad \cos\theta = \frac{\text{adj}}{\text{hyp}} \qquad \tan\theta = \frac{\text{opp}}{\text{adj}}$$

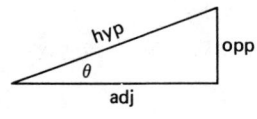

$$1 \text{ rad} = \frac{360}{2\pi} \text{ deg} \approx 57°$$

For any $\theta$, $\sin^2\theta + \cos^2\theta = 1$

$\sin 0° = \cos 90° = 0$

$\sin 30° = \cos 60° = \frac{1}{2} = 0.500$

$\sin 45° = \cos 45° = \frac{1}{\sqrt{2}} = 0.707$

$\sin 60° = \cos 30° = \frac{\sqrt{3}}{2} = 0.866$

$\sin 90° = \cos 0° = 1.000$

## VECTORS

Scalar (dot) product: $\mathbf{A} \cdot \mathbf{B} = |A||B|\cos\theta$
Vector (cross) product: $\mathbf{C} = (\mathbf{A} \times \mathbf{B}) = -(\mathbf{B} \times \mathbf{A})$
$|C| = |A||B|\sin\theta$

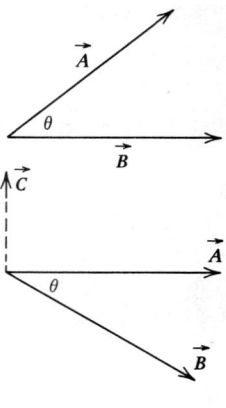

## UNIT CONVERSIONS

Energy:
$1.6 \times 10^{-19}$ J = 1 eV   $\qquad$ 1 J = $6.24 \times 10^{18}$ eV

4.18 J = 1 cal

$3.6 \times 10^6$ J = 1 kWh = $0.86 \times 10^6$ cal

1 J = $9.48 \times 10^{-4}$ BTU (British thermal unit)

1 kg = $9.0 \times 10^{16}$ J/$c^2$

1 eV/molecule = 23 kcal/mol

Length:
1 m = 39.37 in. = 3.28 ft

1 in. = 2.54 cm

1 mile = 5280 ft = 1609 m = 1.6 km

1 light-year = $9.46 \times 10^{15}$ m

Speed:
1 m/s = 3.28 ft/s = 3.6 km/h = 2.24 mph

88 ft/s = 60 mph = 96 km/h = 26.8 m/s

Force:
1 N = 0.225 lb   $\qquad$ 1 lb = 4.45 N

The weight of 1 kg on the earth's surface is 9.8 N = 2.2 lb

Power:
1 kW = 1.34 horsepower   $\qquad$ 1 hp = 0.75 kW

# APPENDIX 2
# GREEK LETTERS AND THEIR COMMON USE IN PHYSICS

| | | |
|---|---|---|
| $\alpha$ | alpha (alfa) | Sometimes used for angle |
| | | Angular acceleration in rad/s$^2$ |
| | | Resistivity coefficient of temperature |
| $\beta$ | beta (bāta) | Sometimes used for angle |
| | | Coefficient of volume expansion |
| $\gamma$ | gamma | Ratio of specific heats of gases |
| | | Surface tension |
| | | Factor in Lorentz transformation |
| $\delta$ | delta (lower case) | A small quantity |
| $\Delta$ | delta (capital) | The change in a quantity |
| $\varepsilon$ | epsilon | Permittivity |
| $\zeta$ | zeta (zāta) | |
| $\eta$ | eta (āta) | Viscosity |
| $\theta$ | theta (thāta) | Usually used for angle |
| $\iota$ | iota | |
| $\kappa$ | kappa | Dielectric constant |

| | | |
|---|---|---|
| λ | lambda | Wavelength |
| | | Linear charge density |
| | | Exponential time constant |
| μ | mu (mew) | Permeability |
| | | Coefficient of friction |
| | | Muon—a heavy electron |
| | | Micro—prefix for $10^{-6}$ |
| ν | nu (noo) | Sometimes used for frequency |
| | | Symbol for neutrino |
| ξ | xi | |
| o | omicron | |
| π | pi (pie) | Ratio of circumference to diameter of circle |
| | | Pion—one of the mesons |
| ρ | rho (rō) | Radius to special point (e.g., to center of mass) |
| | | Mass density |
| | | Resistivity |
| σ | sigma (lower case) | Area charge density |
| Σ | sigma (capital) | Summation |
| τ | tau (taw) | Torque |
| | | Exponential time constant |
| υ | upsilon | |
| φ | phi (fī) (lower case) | Usually used for angle |
| Φ | phi (capital) | Flux (number of lines of force) |

| | | |
|---|---|---|
| χ | chi (kī) | Susceptibility |
| ψ | psi (sī) | probability function in quantum mechanics |
| ω | omega (ōmāga) (lower case) | Angular frequency in rad/s |
| Ω | omega (capital) | Symbol for ohm (electrical resistance) |

## OTHER SPECIAL LETTERS

$\hbar = \dfrac{h}{2\pi}$    $\hbar$-bar is Planck's constant divided by $2\pi$.

ℰ is used for emf, measured in volts (V).

# APPENDIX 3
# THE INTERNATIONAL SYSTEM OF UNITS (SI)

All the sciences, as well as engineering and our daily life, are bedeviled with different units of measure. For instance, lengths can be measured in feet, inches, yards, meters, furlongs, spans, fathoms, ells, miles, kilometers, rods, light-years, and parsecs. For almost 200 years, scientific organizations, sponsored by national governments, have been trying to set up unified standards. The current agreement has resulted in the SI (Système International) units. These are basically the metric units, and have been adopted by most nations. They are rapidly becoming standard in all technical work, including all new physics texts. The units are divided into base, supplementary, and derived units. Note that in general the names, which are often the names of scientists, are uncapitalized. However, the symbols derived from a person's name are capitalized. Thus the unit of energy is J, standing for joule.

## SI BASE AND SUPPLEMENTARY UNITS

|  | Quantity | Unit Name | Unit Symbol |
|---|---|---|---|
| SI base units | Length | meter | m |
|  | Mass | kilogram | kg |
|  | Time | second | s |
|  | Electric current | ampere | A |
|  | Thermodynamic temperature | kelvin | K |
|  | Amount of substance | mole | mol |
|  | Luminous intensity | candela | cd |
| SI supplementary units | Plane angle | radian | rad |
|  | Solid angle | steradian | sr |

## SI DERIVED UNITS WITH SPECIAL NAMES

| Quantity | SI Unit Name | Symbol | Expression in Terms of Other Units | Expression in Terms of SI Base Units |
|---|---|---|---|---|
| Frequency | hertz | Hz | | $s^{-1}$ |
| Force | newton | N | | $m \cdot kg \cdot s^{-2}$ |
| Pressure, stress | pascal | Pa | $N/m^2$ | $m^{-1} \cdot kg \cdot s^{-2}$ |
| Energy, work, quantity of heat | joule | J | $N \cdot m$ | $m^2 \cdot kg \cdot s^{-2}$ |
| Power, radiant flux | watt | W | $J/s$ | $m^2 \cdot kg \cdot s^{-3}$ |
| Quantity of electricity, electrical charge | coulomb | C | $A \cdot s$ | $s \cdot A$ |
| Electric potential, potential difference, electromotive force | volt | V | $W/A$ | $m^2 \cdot kg \cdot s^{-3} \cdot A^{-1}$ |
| Capacitance | farad | F | $C/V$ | $m^{-2} \cdot kg^{-1} \cdot s^4 \cdot A^2$ |
| Electric resistance | ohm | Ω | $V/A$ | $m^2 \cdot kg \cdot s^{-3} \cdot A^{-2}$ |
| Conductance | siemens | S | $A/V$ | $m^{-2} \cdot kg^{-1} \cdot s^3 \cdot A^2$ |
| Magnetic flux | weber | Wb | $V \cdot s$ | $m^2 \cdot kg \cdot s^{-2} \cdot A^{-1}$ |
| Magnetic flux density | tesla | T | $Wb/m^2$ | $kg \cdot s^{-2} \cdot A^{-1}$ |
| Inductance | henry | H | $Wb/A$ | $m^2 \cdot kg \cdot s^{-2} \cdot A^{-2}$ |
| Celsius temperature | degree Celsius | °C | | K |
| Luminous flux | lumen | lm | | $cd \cdot sr$ |
| Illuminance | lux | lx | $lm/m^2$ | $m^{-2} \cdot cd \cdot sr$ |
| Activity (of a radionuclide) | becquerel | Bq | | $s^{-1}$ |
| Absorbed dose, specific energy imparted, kerma, absorbed dose index | gray | Gy | $J/kg$ | $m^2 \cdot s^{-2}$ |
| Dose equivalent, dose equivalent index | sievert | Sv | $J/kg$ | $m^2 \cdot s^{-2}$ |

## SI PREFIXES

| Factor | Prefix | Symbol | Factor | Prefix | Symbol |
|---|---|---|---|---|---|
| $10^{18}$ | exa | E | $10^{-1}$ | deci | d |
| $10^{15}$ | peta | P | $10^{-2}$ | centi | c |
| $10^{12}$ | tera | T | $10^{-3}$ | milli | m |
| $10^{9}$ | giga | G | $10^{-6}$ | micro | $\mu$ |
| $10^{6}$ | mega | M | $10^{-9}$ | nano | n |
| $10^{3}$ | kilo | k | $10^{-12}$ | pico | p |
| $10^{2}$ | hecto | h | $10^{-15}$ | femto | f |
| $10^{1}$ | deka | da | $10^{-18}$ | atto | a |

# INDEX

Abscissa, 39
Absolute error, 12, 13, 19, 22
Acidity, 188
Acceleration, 82, 105
Accuracy, 15, 16
Acute angle, 51, 52
Algebraic, slope derivations, 81, 92, 99, 162
Angle, 46
Approximations, 2, 53, 66, 193
Areas, 64, 77
Area "under curve," 85, 124
Argument of function, 149, 151–153
Axes, 62

Base of exponential, 165–167
Base of logarithms, 176, 177
Binomial expansion, 192

Circular motion, 148
Collisions and momentum, 127
Complement of angle, 51
Components, vector, 112, 116, 127
Conservation:
   of energy, 136
   of momentum, 126–128
Cosine, 48, 57, 63
Cross product, 139–143
Cylinders, 61
Cubic, 90

Data "points," 42
Data "regions," 43
Decibels (dB), 189
Degrees, 47
Delta, 71
Density, 27, 28, 29
Diagrams, 59–65
Dimensions, 34, 35
Displacement, 111
Dot product, 133–138

Effective force, 133
Energy, 132–137
   kinetic, 135
Equilateral triangle, 49
Error, 12, 15, 19, 22
Error bars, 43

Exponential decay, 168
Exponential function, 160–171
   slope, 161, 162
Exponential growth, 164
Exponentials and logs, 178, 179
Extrapolation, 73

Factors of −, 6
Fermi, Enrico, 5
Fermi problems, 6, 7, 8
First approximation, 2, 66
Flux, 138
Force, 76, 104
   effective, 133
Force addition, 115, 116
Force and momentum, 123–126
Fractional error, 19, 23

Geometry, 59–65
Graphs, 37–45
   log, 180–187
Gravitational force, 26, 104
Gravity, acceleration, 78, 80, 82, 84, 105
Greek letters, 195

Half-life, 184
Hooke's law, 76, 104
Hypotenuse, 48

Impulse, 124
Integration, 86
Intensity of sound, 189
International System of Units, 198
Isosceles triangle, 50
Inverse functions, 98
Inverse square effects, 99

Joule, 133, 136

Kilogram, 26
Kinetic energy, 135

Law of cosines, 57, 63
Law of sines, 57, 64
Length measurements, 17, 18
Lever arm, 140
Linear function, 67, 69

Linearity, 69, 71
Logarithms, 172–191
　bases, 176, 177
　calculating, 173, 174
　and exponentials, 178, 179
　graphs, 180–187
　multiplication, 175
　natural base, 177
　slope, 180
Log-log graphs, 184–187

Mass, 26
Meter, 17
Mistakes, 14
Momentum, 123–131
Motion energy, 135
Motion graphs, 86

Negative exponents, 98
Newton, force unit, 76
Newton's law, gravitation, 98, 104
Newton's second law, 104, 107

Octant of sphere, 61
Ordinate, 39

Pace, 18, 19
Parallax, 15, 16
Pascal, 106
Pendulum, 94, 145–147
Percentage, 20, 21
Percentage error, 12, 13, 19, 23
pH (potency of hydrogen), 188
Phase, 153–155
Piano tuners, 5
Pound, 26, 76
Power, 137
Power functions, 66–103
　comparison, 100
Power-of-ten, 1
Precision, 14, 16
Pressure, 105
Propagation of errors, 22
Proportionality, 67, 68
Proton, 1, 2
Pseudovector, 142
Pulse, human, 30
Pythagorean theorem, 49, 50, 57
　proof, 63

Quadratic formula, 192
Quadratic function, 77

Radian, 47, 167
Radioactivity, 168, 180
Reflex, human, 32
Relative error, 19, 20
Relativistic velocity addition, 118

Resolving, vector, 112, 115
Resultant, vector, 112
Right-hand screw rule, 62, 140
Right triangle, 48, 51, 52, 60

Scalar product, 133–138
Scale units, 41
Scientific notation, 1
Seat belts, 125, 126
Second, 30, 31
Semilog graphs, 180–184
Significant figures, 9, 10, 11
Similar triangle, 52
Sine, 48, 57, 64
Sine curve, 50, 147
　slope, 157
Sinusoidal functions, 46, 145–159
SI units, 17, 198–200
Slope, algebraic, 81, 92, 99, 162
Slope of graph, 70, 73, 79, 155, 161, 180
Small angle approximations, 53, 55
Sources of error, 15, 16
Speed, 72, 79
Spheres, 61
Springs, 76, 77
Square root function, 94
Square roots, 192
Standard deviation, 19
Strength tests, 105
Surveying, 54

Tangent, 48, 51
Terminal speed, 82
Time, 30
Torque, 139
Triangle, 48, 51, 52, 60
Trig identities and relationships, 56, 57, 193
Trigonometry, 46–58

Uncertainty, 12, 13
Unit angle, 46
Unit conversions, 194
Units, 34, 35, 36
Universe, 1, 2, 172
Universe, map, 173

Vector product, 139–143
Vectors:
　addition, 111–122
　cross product, 139–143
　dot product, 133–138
Velocity, 72, 79, 119
Velocity addition, 118–120
Volume, 28, 29, 64

Weight, 26, 104
Work, 132, 133